ing, the identity of the Duke de Jarmany,
the horse-stealing subplot. The author
then discusses this in play in relation to
other relevant works in the Shakespearean canon and in terms of the dramatic
modes of the day. He further gives a
reconstruction of the manner in which
Shakespeare may have written the play.

This is one of the few book-length
studies of the *Merry Wives* ever undertaken, and the only one in recent years
to employ historical methods. Both the
serious student of Shakespeare and the
informed general reader will find it an interesting, useful, and informative guide.
It is, in the opinion of G. E. Bentley, "the
best balanced and most comprehensive
discussion of all the major problems of
the play."

This book may be kept
Twenty-one days
A fine of $0.20 per day will
be charged for each day it is overdue

S.

MERRY

SHAKESPEARE'S

MERRY WIVES OF WINDSOR

BY WILLIAM GREEN

PRINCETON, NEW JERSEY
PRINCETON UNIVERSITY PRESS
1962

To My Mother

PREFACE

I N PREPARING this manuscript, I have adhered to the following procedures:

All quotations from the folio text of the *Merry Wives* have been taken from the Yale photographic facsimile of the First Folio prepared by Helge Kökeritz and Charles Tyler Prouty, with lineation from the Globe edition. Quotations from the First Quarto (1602) have been taken from the collotype facsimile of Q1 published in 1939 by the Shakespeare Association in its Shakespeare Quarto Facsimile series. Since Q1 bears neither pagination nor scene division, I have used the continuous lineation employed by W. W. Greg in his 1910 facsimile edition of the Quarto. In spite of the archaisms in these texts, they are more valuable in a study concerned with the composition of the play than a modernized edition.

In quoting from other early printed texts and manuscripts, I have retained original spelling and punctuation, but have silently expanded endings and abbreviations. Superscriptions have been brought down to the line. In documenting records from the Public Record Office, where both Audit Office and Pipe Office entries are extant and identical, I have cited only the Audit Office copy.

Latin passages have been silently rendered into English inasmuch as for purposes of this study I have not deemed it necessary to retain the original Latin.

Unless the compounded form for noting the year appears, e.g., January 18, 1601/02, all dates have been normalized according to modern usage. When appropriate, I have indicated old style and new style dates; otherwise all days of the month are recorded according to normal English procedure of the Elizabethan period.

All references to and quotations from the statutes of the Order of the Garter have been taken from Henry VIII's revised English version, which was in force throughout Elizabeth's reign. These statutes exist in an equally official Latin version which has some differences in the numbering of articles and has a completely reworded Article XVIII.

I wish to acknowledge my indebtedness to those who helped me with this study of the *Merry Wives*. Without the generosity of the University of Manchester in awarding me a fellowship, the archival research could not have been undertaken. For aid in gaining access to, examining, and interpreting manuscripts I am indebted to The Hon. Sir George Bellew, Garter Principal King of Arms; The Rt. Rev. Eric Hamilton, Dean of Windsor and Registrar of the Most Noble Order of the Garter; Mr. Maurice Bond, Hon. Custodian of the Muniments at Windsor Castle; Mr. E. K. Timings of the Public Record Office; Dr. M. Tyson and Dr. G. Aylmer of the University of Manchester. For criticism and advice during the preparation of this manuscript in its original form as a doctoral dissertation, I am grateful to the late Professor H. B. Charlton of the University of Manchester; and to Professors Ernest Brennecke, Oscar J. Campbell, S. F. Johnson, Edward Le Comte, and Garrett Mattingly of Columbia University. It was Professor Campbell who first encouraged me to proceed with my investigations. I have also profited greatly from the reading my friend Professor Donald A. Sears of Upsala College gave the manuscript in preparation for the press. For assistance with translation problems, I am indebted to Professor Konrad Gries of Queens College. For countless services I thank the librarians at the Bodleian, British Museum, Dulwich College, John Rylands, Manchester Central, and University of Manchester libraries in England; and at the New York Public and Queens College libraries in the United States. I further wish to thank the staff of the Princeton University Press, particularly Mrs. James Holly Hanford, for their kindnesses and the Trustees of the Ford Foundation for the partial grant which made publication of this book possible. My deepest thanks to my wife, Marguerite, for her help and support.

Permission to reproduce illustrations and/or extracts from the manuscript collections of the Bodleian Library and the British Museum is gratefully acknowledged. "Unpublished Crown copyright material in the Public Record Office, London, has been reproduced by permission of the Controller of H. M. Stationery Office." The Bodley Head Ltd. has kindly granted permission to reproduce passages from T. H. Nash's translation of Breuning von Buchenbach's account of his 1595 ambassadorial mission to England which appears in *Queen Elizabeth and Some Foreigners. . .* , edited by Victor von Klarwill. Quotations from *Shakespeare versus Shallow* are reprinted with the permission of the author, Leslie Hotson.

Lastly, I can acknowledge only in a general way my indebtedness to the many Shakespearean scholars whose writings have influenced me over the years.

W.G.

New York, New York
October 1961

CONTENTS

ILLUSTRATIONS

10. Arms within a Garter of Queen Elizabeth I; William Brooke, Lord Cobham; George Carey, Lord Hunsdon; Frederick, Duke of Württemberg, from the 1780 series of hand-painted authentic reproductions of the coats of arms of all Knights of the Garter to 1771 by Joseph Edmondson, Mowbray Herald Extraordinary; B.M. King's MS. 408, fols. 3, 33, and 49; King's MS. 409, fol. 9.

(The photographs for Nos. 1, 2, 4-11 are by the British Museum Photographic Service. The photograph for No. 3 is by the New York Public Library Photographic Service.)

SHAKESPEARE'S

MERRY WIVES OF WINDSOR

ABBREVIATIONS

Add.	Additional Manuscripts
Ashm.	Ashmolean Manuscripts
B.M.	British Museum
Bodl.	Bodleian Library
Coll. Arms.	College of Arms
Cott.	Cottonian Manuscripts
Harl.	Harleian Manuscripts
H.M.C.	Historical Manuscripts Commission
Lansd.	Lansdowne Manuscripts
Lib.	*The Library*
MLR	*Modern Language Review*
N&Q	*Notes and Queries*
OED	*Oxford English Dictionary*
PMLA	*Publications of the Modern Language Association of America*
PQ	*Philological Quarterly*
P.R.O.	Public Record Office
RES	*The Review of English Studies*
SAB	*Shakespeare Association Bulletin*
SP	*Studies in Philology*
SQ	*Shakespeare Quarterly*
TLS	*Times Literary Supplement*

1. WINDSOR
2. MAIDENHEAD
3. READING
4. COLEBROOKE

". . . there is three Cozen-Iermans, that has cozend all the *Hosts* of *Readins*, of *Maidenhead*; of *Cole-brooke*, of horses and money. . . ."

1. BERKSHIRE ca. 1576, by Christopher Saxton.

2. WINDSOR CASTLE and environs, 1607, by John Norden.

1. THE GARTER INN. "Do's he lye at the Garter?"

2. DATCHET MEAD. ". . . send him by your two men to *Datchet*-Meade."

3. "HOG HOLE." (The Ditch.) ". . . take this basket . . . and . . . empty it in the muddie ditch, close by the Thames side."

4. DATCHET LANE. ". . . a couple of *Fords* knaues . . . were cald forth by their Mistris, to carry mee in the name of foule Cloathes to *Datchet-lane*. . . ."

5. THE FIELDS. ". . . I will bring the Doctor about by the Fields. . . ." (Dr. Caius was waiting somewhere on the north side of the castle.)

6. THE ROAD TO FROGMORE. ". . . Mr. Ghuest, and M. *Page*, & eeke Caualeiro *Slender*, goe you through the Towne to *Frogmore*."

7. THE CASTLE DITCH. "Come, come: wee'll couch i'th Castle-ditch, till we see the light of our Fairies."

8. HERNE'S OAK. "Bee you in the Parke about midnight, at Hernes-Oake, and you shall see wonders."

9. THE DELL (Saw Pit). "Let them [the Fairies] from forth a saw-pit rush at once/ With some diffused song. . . . They are all couch'd in a pit hard by Hernes Oake, with obscur'd Lights. . . ."

3. The Choir of St. George's Chapel, Windsor, as it was in Shakespeare's time.

1. The High Altar
2. Knights Companions Stalls
3. Canons Stalls in absence of the knights
4. Petty Canons and Vicars Stalls
5. Choristers Seats
6. Alms Knights Chaires

7. Knights Companions Banners
8. Their Helmes, Crests, and Swords
9. Plates of their Armes and Styles
10. Queenes Closet
11. Closet for Ladies
12. East Window of the Choire

The seuerall Chaires of Order, looke you scowre
With iuyce of Balme; and euery precious flowre,
Each faire Instalment,[2] Coate,[9] and seu'rall Crest,[8]
With loyall Blazon,[7] euermore be blest.

(The figures in brackets identify the locations in the above picture.)

INTRODUCTION

ISTORY," Hume has written, "being a collection of facts which are multiplying without end, is obliged to adopt arts of abridgment—to retain the more material events, and to drop all the minute circumstances, which are only interesting during the time, or to the persons engaged in the transactions." This study is about one of those minute circumstances: the composition of *The Merry Wives of Windsor*.

The *Merry Wives* is one of the most individualistic plays that Shakespeare wrote. It is his sole play set in its entirety in an Elizabethan milieu. It contains a set of characters also found in *1* and *2 Henry IV* and *Henry V* but who have no biographical links to their namesakes in the history plays. Prose is the medium used almost exclusively throughout the text. Textual specialists have classified the First Quarto as a "bad" quarto. The Folio text, on the basis of certain scribal characteristics in it, belongs among the group of plays thought to have been printed from Ralph Crane transcripts. No definitive source has ever been discovered for the play. And the matter of its date has not been settled.

Theater audiences seem less sensitive to being confronted with a freak than scholars, for the *Merry Wives* has had a long and popular stage history. In performance, the play becomes a rollicking farce. It has won additional enthusiastic followers through its musical versions: Nicolai's *The Merry Wives of Windsor*, Verdi's *Falstaff*, and Vaughan Williams' *Sir John in Love*. Since the *Merry Wives* was created for the stage, it is by its stage durability that we must give it its final due.

Yet, all the aforementioned oddities do invite the scholar to a feast of literary investigation. Surprisingly, few—and those mainly textual critics—have partaken of the morsels offered. In this study, as stated above, I restrict myself primarily to one aspect of the play: the events surrounding its composition and the manner by which those events shaped the text of the play from its initial performance through to the 1623 Folio printing.

Impetus for the present investigation stems from the theory of dating for the *Merry Wives* advanced by Leslie Hotson in *Shakespeare versus Shallow* (1931). Using both external and internal evidence, Hotson concluded that the play was first performed at the Feast of the Order of the Garter celebrated on St. George's Day, 1597. However, interested in other matters in *Shakespeare versus Shallow* (namely the identification of Justice Shallow with the Surrey Justice of the Peace, William Gardiner), Hotson did not examine all the ramifications of his theory. In fact, he paid only a minimal amount of attention to it—sufficient, of course, for his purpose.

I aim to reexamine the Hotson dating theory, which stands apart from the major thesis of his book, in an attempt to prove its validity. For the date is inextricably bound up with the composition of the play in the sense that, as I shall try to show, Shakespeare was not suddenly moved by inspiration to write a comedy of contemporary village life centered around Falstaff; instead he wrote the play to order. And the postulated date of the first performance provides the key to unlocking the secrets of why Shakespeare wrote the play and how he proceeded to do it.

The events surrounding the composition of the *Merry Wives* have long been relegated to the dead past as too trivial for history to have kept alive. Yet extant documents make it possible to re-create them. In some cases there appears to be direct evidence; in others, hearsay and news spread by court gossips furnish grounds for inference. What a study of these materials permits us to do is to re-enter the Elizabethan world and to discover by savoring the contemporary atmosphere how closely situations and allusions in the *Merry Wives* reflect current events. Having taken such a step, we place ourselves in a better position to conceive the design of the play, even to understand why the script is an imperfect one. Falstaff in a romantic entanglement; the presence of his cronies from the *Henry* plays; allusions to the Order of the Garter; the contemporary Windsor setting; the Brooke-Broome name variant between the Quarto and Folio versions; the strange, unintegrated horse-stealing subplot—all these items become comprehensible when we learn what the play meant to the Elizabethan, whether he was a courtier or commoner.

To the Elizabethan, the *Merry Wives* was Shakespeare's Garter play. This hypothesis, of course, calls for substantiation. Accordingly, in the plan of this book I first present a picture of the elements in the script which fix the play as a Garter play. Then I establish the historical background to support my thesis. Following this, I turn to matters of text. Here I do not try to duplicate the able detailed textual analyses that have been made of the play, but primarily I discuss the relationship of the Quarto and Folio versions to one another in an attempt to recapture the script as presented at the postulated 1597 production and to explain what I believe happened to it in the intervening years before it was printed in the First Folio. Since this study of text takes into consideration matters of topical significance, I devote separate chapters to examination of the Brooke-Broome variant reading, the identity of the Duke de Jarmany of the horse-stealing subplot, and to the subplot itself. In the latter portions of this work I place the *Merry Wives* in the contemporary literary setting from which it emerged; that is, I discuss it in terms of other relevant plays in Shakespeare's canon and in terms of dramatic modes of the day to show that there is nothing incompatible between Shakespeare's dramatic output as a whole and a 1597 date for the comedy. In this connection, I have deliberately placed in an appendix discussion of the absence of the *Merry Wives* from Meres's list of Shakespeare's plays in the *Palladis Tamia*, for scholarly research more and more has revealed that Meres cannot be accepted as an unimpeachable witness for evidence about the Elizabethan literary world; and I believe that this point of view particularly holds true for the *Merry Wives*. Lastly, I attempt a hypothetical reconstruction of the manner in which Shakespeare wrote the play.

Throughout this treatise I have tried to steer a straight course between the Scylla of writing historical fiction and Charybdis of equating inferences with facts. For above all, I realize that until a document dated late April 1597, or thereabouts, may one day be unearthed in which some chronicler states, "Saw a newe plaie of Ffalstoffe in Loue which brought grete mirthe to th'assemblie gathered at court on St. Georges daie," the case herein presented remains a circumstantial one.

CHAPTER I + THE *MERRY WIVES* AND THE ORDER OF THE GARTER

BASICALLY the *Merry Wives* is a play of English village life. As such, it colorfully portrays the activities of some of the more prosperous inhabitants of an Elizabethan country town. But why should Shakespeare, who knew Warwickshire so intimately, have set his play in Windsor? Why not have written *The Merry Wives of Stratford* or *The Merry Wives of Banbury* or what you will? Why not indeed—unless Windsor had special significance for the Elizabethans.

Windsor has never been just another picturesque town in the Thames valley area. Since the time of William the Conqueror, it has continuously been the site of a royal residence. Windsor Castle, standing majestically on a chalk cliff high above the river Thames, was to the Elizabethan—and still is to the present-day Englishman—a symbol of British monarchy.

And since the fourteenth century, Windsor Castle has held a further association in the public mind, one aptly brought forth by Shakespeare's contemporary Michael Drayton in some lines from his *Poly-Olbion*. In "Song XV" Drayton climaxes his description of the flow of the Thames through Windsor with:

> Then, hand in hand, her *Tames* the Forrest softly brings,
> To that supreamest place of the great English Kings,
> The *Garters* Royall seate, from him who did advance
> That Princely Order first, our first that conquered *France*;

The Temple of *Saint George,* wheras [*sic*] his
honored Knights,
Upon his hallowed day, observe their ancient
rites.[1]

Windsor, then, carries with it the distinction of being
the home of the Order of the Garter. Stow mentions the
connection in his description of Windsor Castle (ca.
1572),[2] and Camden in his *Brittania. . .* (1586) includes a
poem entitled "The Marriage of Tame and Isis" which
contains a passage extolling Windsor and the Order of the
Garter.[3] Not only was the link known among Englishmen,
but among foreigners. George Braun and Francis
Hohenberg in the 1575 edition of their travel book
Civitates Orbis Terrarum note of Windsor that "this
Castle is most famous for being the home of the royal
family, for having magnificent royal tombs, and for the
ceremonies pertaining to the Knights of the Garter. . . .
Now the ceremony of this Order is held yearly at Windsor
on a given day sacred to St. George, tutelar Saint of the
Knights, with the King presiding; and it is the custom that
the members hang up their helmets and shields, with their
escutcheons emblazoned thereon, in a conspicuous part of

---

[1] *The Works of Michael Drayton,* ed. J. William Hebel (Oxford, 1933),
Vol. IV. In his note to line 315, John Selden comments: "I cannot but
remember the institution . . . of his most honorable Order, dedicated to
S. George . . . it is yeerly at this place celebrated by that Noble companie
of XXVI. Whether the cause were upon the word of *Garter* given in the
French wars among the *English,* or upon the Queens, or Countes of
Salisburies Garter fallen from her leg, or upon different & more ancient
Original whatsoever, know cleerly (without unlimited affection of your
Countries glorie) that it exceeds in Majestie, honor, and fame, all Chival-
rous Orders in the world; and (excepting those . . . which were more
Religious then Military) hath precedence of Antiquity before the eldest
rank of honor, of that kind any where established." The explanatory notes
for the First Part of *Poly-Olbion* (Songs I-XVIII) were specially written by
the learned John Selden at the request of Drayton, his close friend.

[2] B.M. Harl. MS. 367, fols. 13-13v; reprinted in W. H. St. John Hope,
Windsor Castle: an Architectural History (London, 1913), I, 280.

[3] 4th ed. [1607], trans. Edmund Gibson (London, 1772), I, 231-232.

the chapel."[4] Inasmuch as the *Civitates* appeared in three languages—Latin, German, and French—and in several editions during the late sixteenth and into the seventeenth century, we can get some idea of how widespread was the knowledge of Windsor as the home of the Order of the Garter. (Provided, of course, that the books did not become mere decorative volumes on the shelves.) Even Paul Hentzner, a German visitor to the Castle in 1598, used the *Civitates* account, unacknowledged, as the source for much of the description of Windsor appearing in his journal.[5] Hence we see from these various accounts that Windsor and the Order of the Garter were commingled in the minds of the Elizabethans and some of their foreign contemporaries. By selecting the town for the locale of the *Merry Wives*, then, Shakespeare deliberately chose a setting that would evoke an Order of the Garter association for his audience.

Not only by setting but by actual dialogue does Shakespeare establish his Garter motif in the *Merry Wives*. In V.v.59-77, the Fairy Queen indicates that some ceremonial related to the Order is to take place at the castle:

> About, about:
> Search Windsor Castle (Elues) within, and out,
> Strew good lucke (Ouphes) on euery sacred
>   roome,
> That it may stand till the perpetuall doome,
> In state as wholsome, as in state 'tis fit,
> Worthy the Owner, and the Owner it.
> The seuerall Chaires of Order, looke you scowre
> With iuyce of Balme; and euery precious flowre,

---

[4] *Civitates Orbis Terrarum* (Brussels, 1572-1617) is the running title of the series of six volumes. This passage appears in Vol. II, entitled *De Praecipuis Totius Universi Urbibus, Liber Secundus*, p. 2.

[5] See *England as Seen by Foreigners. . .* , ed. W. B. Rye (London, 1865), pp. 198-199.

Each faire Instalment, Coate, and seu'rall Crest,
With loyall Blazon, euermore be blest.
And Nightly-meadow-Fairies, looke you sing
Like to the *Garters*-Compasse, in a ring,
Th'expressure that it beares: Greene let it be,
Mote [*sic*] fertile-fresh then all the Field to see:
And, *Hony Soit Qui Mal-y-Pence*, write
In Emrold-tuffes, Flowres purple, blew, and
   white,
Like Saphire-pearle, and rich embroiderie,
Buckled below faire Knight-hoods bending knee;
Fairies vse Flowres for their characterie.

These laudatory lines—completely extraneous to the plot
of the play—are even preceded by what appears to be a
direct tribute to Queen Elizabeth:

Cricket, to Windsor-chimnies shalt thou leape;
Where fires thou find'st vnrak'd, and hearths
   vnswept,
There pinch the Maids as blew as Bill-berry,
Our radiant Queene, hates Sluts, and Sluttery.

> (V.v.47-50)

   Such passages are too carefully worked out to be ignored
in a play which is as much "a thing of threads and patches"
as is the *Merry Wives*. Even the hues of the flowers selected
by the fairies for their charactery reflect the colors in the
badge of the Order.
   There is still other internal evidence for linking the
Merry Wives and the Order. In I.iv.54, Dr. Caius says, "*Ie
man voi a le Court la grand affaires.*" A few lines later
(ll. 61-62) he instructs Jack Rugby, "Come, take-a-your
Rapier, and come after my heele to the Court." Upon his
exit, the doctor repeats the command, "*Rugby*, come to the
Court with me" (l. 130). This "grand affair" which Caius

is hurrying to—as will shortly be shown—is the same ceremonial for which the fairies are preparing the castle.

Caius is not the only one on his way to Court, i.e., Windsor Castle. The text also reveals that the town is filling
with courtiers. Their presence is made known by Mistress
Quickly. In the same passage in which she has told Falstaff
what happened to Mistress Ford when the Court last was
present (II.ii.62-81), Quickly reveals (and profitably for
her) that "I had my selfe twentie Angels giuen me this
morning" from some of the new arrivals for her to serve in
her old capacity as intermediary to her mistress. She elaborates no further on the presence of these arrivals.

What is this "grand affair" that is drawing the nobles
and Dr. Caius to the castle? Four lines in the Fairy Queen's
instructions for the preparation of the castle focus attention on it:

> The seuerall Chaires of Order, looke you scowre
> With iuyce of Balme; and euery precious flowre,
> Each faire Instalment, Coate, and seu'rall Crest,
> With loyall Blazon, euermore be blest.
>
> (V.v.65-68)

The Chairs of the Order are the stalls of the Garter
Knights in the choir of St. George's Chapel, and the coat,
crest, and blazon are the achievements which hang above
each Knight's stall. So the fairies are making ready the
Garter's own chapel. For what occasion? The only one possible is an installation of Knights-Elect. This was the sole
Garter ceremonial always solemnized at Windsor and the
only one during which the stalls in St. George's Chapel
were occupied and the achievements handled (those of
new Knights offered up and those of deceased Knights
removed). Moreover, between 1567 and 1603—by decree
of Elizabeth entered in the Chapter proceedings for April
23, 1567—no other type of Garter rite except the installa-

tion could be celebrated at Windsor. Accordingly, it is my belief that the "grand affair" of Dr. Caius represents a Garter installation. (See fig. 3.)

And there is nothing unusual about Caius receiving an invitation to the "grand affair." The doctor himself relates that he numbers the nobility among his patients (II.iii.95-97). Mistress Page confirms this with her observation, "The Doctor is well monied, and his friends/Potent at Court" (IV.iv.88-89). Caius is undoubtedly not only the leading (and perhaps sole) doctor in Windsor, but also the official physician to the inhabitants of the castle. Extant lists of categories of individuals called upon to be in attendance at Garter functions reveal that among those so ordered have been physicians, apothecaries, and surgeons. There is, accordingly, no need to question whether Dr. Caius could get a summons to the "grand affair" at the castle.

In retrospect, therefore, we see that the direct Garter allusions in the play, the remarks of Caius, Quickly's note about noblemen in town, the Windsor setting—all seemingly isolated fragments in the play—merge to form a pattern readily intelligible to an Elizabethan audience. To an Elizabethan it was apparent that Shakespeare in the *Merry Wives* had depicted contemporary Windsor at the time of an Order of the Garter installation.

The concept of honoring the Order through specially composed literary works was not new with Shakespeare. The earliest example I have found dates from 1488, and consists of a series of verses by an unknown poet presented at the Feast of St. George which Henry VII had celebrated at Windsor that year.

In Shakespeare's own time, one William Teshe (or Tashe), in 1582, composed an untitled Garter poem dedicated to the Earl of Bedford.[6] It is set at Elizabeth's court

---

[6] It has been printed from B.M. Harl. MS. 3437 in Nicholas Harris Nicolas, *History of the Orders of Knighthood of the British Empire* . . . (London, 1841), II, Appendix H, xxix-xxxviii.

on some great occasion at which the Queen was present. Though unspecified, the occasion could be a Grand Feast. Teshe divides his poem into sections, each devoted to extolling the virtues of one of the following: Queen Elizabeth, the Chancellor of the Order, and the sixteen English Knights of the Garter in 1582. Teshe builds his laudatory verses around the mottoes of these individuals. Of particular interest is the interweaving of the Garter motto (as in V.v. of the *Merry Wives*) in the opening section dedicated to the Queen and headed accordingly, "The Royal Arms, Within the Garter and the Motto, 'Dieu et mon Droit.'" Within this section appears the stanza:

> Shame to the mynde that meanes, (quod Shee,) amisse,
> Whereby was seen her mynde did meane no ill.
> Lo, thus my Lordes our verdict geven up is.
> Lett them do well that looke for our good will
> *A qui mal pense a luy tout honi soit*;
> And for myself, *Mon Dieu & seul mon droit.*

And the poem closes:

> Thus when eche one had geven up his Bende,
> Her Highnes rose from forth her Cheare of State,
> 'I thancke you all, quod Shee, and for an ende,
> Long maie your dayes with myne rest fortunate.'
> Wherwith they all made humble reverence then,
> And all th'assembly said therto, 'Amen.'
>
> The Trumpettes blewe, and Heralds lowd did call,
> *'Sortez Seigneurs, chascun a son Logis.'*
> The Nobles rose, and thence departed all
> Them to disrobe, as use and custome is;
> And as the Earle of Bedforde past by,
> 'Nowe, good my Lorde, remember me,' quod I.

As the above verses illustrate, this poem contains certain features in its treatment of Garter material which are also found in the later *Merry Wives*. Its setting is contemporaneous with the events narrated; it employs the device of interweaving the motto of the Order with the content; and, as we shall soon see for the *Merry Wives*, it was composed with a definite member of the Order in mind. This is not to claim that Shakespeare knew or even utilized Teshe's poem, which in Elizabethan times seems to have existed only in manuscript. All we can say is that the Teshe work was in circulation in 1582 and that if Shakespeare had read it at some time, he may have picked up an idea or two.

Even if he lacked knowledge of this earlier poem, Shakespeare most certainly was acquainted with George Peele's "The Honour of the Garter." Peele wrote it in June 1593, as a tribute to the Earl of Northumberland upon that nobleman's election to the Order. "The Honour of the Garter" is a laudatory work to the five new Knights of the Order—the Earls of Northumberland and Worcester, the Lords Burgh and Sheffield, and Sir Francis Knollys. In describing the greatness of the Order and in stressing the honor in becoming a Knight of the Garter, Peele throughout the poem reveals an accurate knowledge of the history of that organization.

At several points in the work—which is set in a meadow by Windsor Castle where the poet falls asleep and dreams of a procession of Knights of the Garter from the House of Fame—Peele interweaves and stresses the motto of the Order as Teshe had done before him and as Shakespeare was to do not long after. In giving the early history, for example, he writes:

> King Edward wistlie looking on them all,
> With Princely hands having that Garter ceazd
> [seized],

From harmlesse hart where honour was engraved,
Bespake in French. . .
Honi Soit Qui mal y pense, quoth he.

(ll. 120-127)

Then several lines later he adds:

. . . but in a twinck me thought
A [the Black Prince] chaungd at once his habite
and his Steede,
And had a Garter as his father had.
Right rich and costly, with embroyderie
Of Pearle and Gold. I could on it discerne,
The Poesie whereof I spake of yore;
And well I wot since this King Edwards dayes,
Our Kings and Queenes about theyr royall Armes,
Have in a Garter borne this Poesie.

(ll. 151-159)

Peele also has the Earl of Bedford repeat the tale about
the attempted theft of his Garter as he slept by a farm-
house. The Earl concludes his account, stating smilingly:

A [the thief] would not (had a understood the
french
Writ on my Garter) dared t'have stolne the same.

(ll. 307-308)

The Earl's story causes the poet to meditate on the Order's
motto:

Thys tale I thought upon, told me for trueth:
The rather for it praisde the poesie,
Ryght grave and honourable that importeth much.
Ill be to him (it sayth) that evill thinkes.

(ll. 309-312)

This device of placing the Garter motto in the verse is
but one of several parallels in handling Garter subject

matter found in "The Honour of the Garter" and the *Merry Wives*. The poem—like the play—is set in Windsor. In both works the Castle, home of the Order, stands in the background; literally in the poem, figuratively in the play (the "grand affair" at the castle which Dr. Caius is preparing for). The theme of Peele's long poem—it runs to about five hundred lines—is echoed in Shakespeare's concise tribute to the Order in V.v. Also, the author of "The Honour of the Garter" was, like Shakespeare, an outstanding poet and playwright of the day. Furthermore, Peele wrote his poem specifically for the occasion of Northumberland's election, even timing it to appear just before the Earl's installation.[7] Thus the work pointed up the coming ceremonies just as I shall shortly try to show that the *Merry Wives* points up the 1597 installation of Lord Hunsdon, patron of Shakespeare's company.

It has not been established whether Northumberland was Peele's patron at the time or whether Peele wrote the poem on speculation, hoping that he would touch a generous nerve in Northumberland. But the Earl responded favorably to the tribute, and on June 23—a few days before his installation—he had delivered "to one Geo. Peele, a poett, as my Lord's liberality, [£] 3"[8]—quite a handsome sum. Surely Peele's poem, with the large reward that accompanied it, must have made some impact in London

---

[7] Peele's dedication is made in the opening lines of the prologue:
　　Plaine is my coate, and humble is my gate,
　　Thrice noble Earle, behold with gentle eyes
　　My wits poore worth: even for your noblesse,
　　(Renowmed [*sic*] Lord, Northumberland's fayre flower).
He ends the poem stating that it is
　　Consecrated purely to your [Northumberland's] noble name,
　　To gratulate to you this honours heigth,
　　As little boyes with flinging up their cappes,
　　Congratulate great Kings and Conquerors.
[8] From the accounts of the Duke of Northumberland printed in the *Historical Manuscripts Commission, Sixth Report of The Royal Commission on Historical Manuscripts* (London, 1877), p. 227.

literary circles. And by 1593 Shakespeare was a member of those circles.

I am not suggesting that Shakespeare became a slavish imitator of Peele. But—let us conjecture—here is Shakespeare the craftsman intent upon preparing a play in about two weeks by royal request, as an old stage tradition relates (a tradition which will be evaluated in a later chapter). He conceives the idea of inserting Garter elements into the play. As he ponders how to go about it, he recollects the Peele poem—written especially for the 1593 induction ceremonial of the Order. Shakespeare obtains a copy of the poem. The very title page begins to give him ideas. It reads in part: "The Honovr of the/Garter. Displaied in a Poeme gratulatorie: Entitled/to the worthie and renowned Earle/of Northumberland./Created Knight of that Order, and installd at/Windsore. Anno Regni Elizabetha. 35./Die Iunij. 26. . . ." A perusal of the poem suggests an approach to him—nothing more. And, as I shall show in depth in the following chapters of this study, Shakespeare begins to create what I call his Garter play— a specially composed work for first presentation in connection with a ceremonial of the Order of the Garter.

We gain some insight into the date of the ceremonial and the reason for presentation on that occasion by a consideration of the time setting of the play. We have already noted that Shakespeare portrayed contemporary Windsor. This is not an assumption, for antiquarians have shown us that Shakespeare depicts Windsor as it was in the 1590's.[9]

A Garter Inn had been located in High Street opposite the castle during the sixteenth and seventeenth centuries. The two points at which Caius and Evans await one another for their duel—Frogmore and the Fields—lay, as in

---

[9] See, for example, Robert Richard Tighe and James Edward Davis, *Annals of Windsor* . . . (London, 1858), I, 666-705; also T. Eustace Harwood, *Windsor Old and New* (London, 1929), pp. 117, 275-276.

the play, at opposite sides of the town (although the exact area of the Fields is somewhat in doubt). The route by which the servants carry Falstaff in the buck-basket from town through Datchet Mead to the Thames actually existed in Elizabethan Windsor. Even the topography of the park setting of the final scene accords with that of the Little Park as it was in Shakespeare's day. The castle ditch from which Page, Shallow, and Slender are to watch for the fairy lights; the pit from which the fairies are to rush upon Falstaff; and Herne's Oak all had actual counterparts in proximity to one another on the park grounds.[10] (See map, fig. 2.)

Not only is it Windsor of the 1590's that Shakespeare presents, but, more precisely, it is probably the Windsor of 1597. The basis for this statement stems from my proposition that the "grand affair" at the Castle represents an installation of new Knights of the Garter. In May 1597, the installation of Lord Hunsdon (Shakespeare's patron) and three other Knights-Elect took place at the Castle. Before that date Windsor had not served as the locale for a Garter ceremonial since June 1593. In fact, the Castle appears to have been used very little in the interval. Lines in the fairy scene tend to bear out this observation: "Cricket, to Windsor-chimnies shalt thou leape;/Where fires thou find'st vnrak'd, and hearths vnswept" (V.v.47-48).

There was, however, one major event which occurred shortly after the 1593 Garter installation. From August through December that same year the Queen held Court at Windsor, during which interval she made her translation

10 The existence of the oak has been attested to by various Windsor residents, although there has never been unanimous agreement as to its precise location. Tighe and Davis are convinced that the original stood in the Little Park on the right of a footpath leading from Windsor to Datchet and that it was cut down in 1796. Charles Knight, who was a native of Windsor, has an interesting little essay on the subject in his *Pictorial Edition of the Works of Shakspere* (London [1839-42]), "Comedies," I, 202-205.

of Boethius' *De Consolatione Philosophiae*.[11] Whether by accident or design (and the truth will never be known), one passage in the play appears to indicate that Shakespeare took cognizance of this royal visit. In II.ii.62-81, Mistress Quickly, in flattering Falstaff over the supposed reception of his letter to Mistress Ford, tells him that "the best Courtier of them all (when the Court lay at *Windsor*) could neuer haue brought her to such a Canarie: yet there has beene Knights, and Lords, and Gentlemen, with their Coaches; I warrant you Coach after Coach. . . ." If Quickly's allusion to "when the Court lay at Windsor" may be interpreted as a literal topical allusion, the year of that occasion has to be 1593. After that autumn stay of the Queen, so far as I can ascertain, Windsor did not serve as the locale for any ceremonial occasion until the Garter installation of 1597.

If, then, the installation in the play delimits the portrayal of contemporary Windsor to 1597, we can advance the proposition that Shakespeare was paying homage to Lord Hunsdon upon his election to the Order of the Garter in that year just as Peele had done to the Earl of Northumberland at the last Garter election and installation in 1593. We shall see in the chapters to come how Shakespeare grafted this salute to Hunsdon and the Order on to the Falstaff-in-love plot of the play, and why he did so.

The fact that it is impossible to make a day-by-day account of preparations for the installation, arrival at Windsor, or actual administering of the oath of fealty from the various Garter passages in the *Merry Wives* is unimportant, for we are dealing with a work of art (albeit a minor and hastily written one), not with the orders of the day of the Queen's commissioner for the installation. Besides, like any good dramatist, Shakespeare stuck to historical ac-

11 John Nichols, ed., *The Progresses . . . of Queen Elizabeth*, new ed. (London, 1823), III, 227, 564.

curacy only so far as he felt necessary. Once his allusions were sufficiently recognizable, he embellished and altered them in any way he saw fit according to his artistic aim. Unfortunately, the passage of time has made obscure to later audiences what was obvious to Elizabethan viewers. But the Elizabethan theatergoer, from the moment he found himself at a play set in Windsor, would have easily comprehended that he was witnessing Shakespeare's Garter play.

CHAPTER II ✦ WHEN THE PLAY WAS FIRST PERFORMED

HE first scholar to take cognizance of the Garter references in V.v. of the *Merry Wives* was Edmund Malone, pioneer of Shakespearean chronology. Malone believed that the lines possibly reflected on the Feast of the Order of the Garter celebrated in July 1603, when James I's Court was at Windsor. He concluded, however, that "the place in which the scene lay, might, it must be owned, have suggested an allusion to it [the Order of the Garter], without any particular or temporary object."[1]

Few scholars followed Malone in investigating the possible relationship between the *Merry Wives* and the Order. E. K. Chambers, for example, briefly toyed with two different theories without coming to any conclusion. In *The Elizabethan Stage*[2] he remarks that the play may have been prepared for the Garter Feast on April 23, 1600. In *William Shakespeare*[3] he suggests a performance before the Knights of the Garter at Windsor about 1601. Indeed, no serious attempt was made to assess the evidence until Leslie Hotson turned his attention to the problem in *Shakespeare versus Shallow*.

Hotson's primary consideration in this book was to

---

[1] *The Plays and Poems of William Shakspeare* (London, 1790), I, Part i, 329. A believer in a "first sketch" theory, Malone had already dated the play in 1601. The Garter reference, along with several others, served him as evidence for revision of the play about 1603. Until the "first sketch" theory finally was disposed of, early scholars were constantly misled in their attempts to date the *Merry Wives*. Their efforts were further confounded by erroneous interpretation and attempted linking of the facts that Count Mompelgard made his visit to England in August 1592, and that *The Jealous Comedy* received its sole performance in January 1593. See below, Chapters VII and X.

[2] *The Elizabethan Stage* (Oxford, 1923), II, 204.

[3] *William Shakespeare* (Oxford, 1930), I, 432, 434-435.

identify Justice Shallow of *2 Henry IV* and the *Merry Wives* with the notorious Surrey Justice William Gardiner and also to identify Abraham Slender of the *Merry Wives* with Gardiner's stepson William Wayte. Hotson's researches in this area led him to attempt a dating of the *Merry Wives* somewhere between November 1596 and November 1597.

Having concluded that Shakespeare deliberately satirized Gardiner and Wayte in the two plays, Hotson proceeded to examine the stage traditions concerning the writing of the *Merry Wives* and the possible occasion which could have prompted the appearance of such a hurriedly composed piece.[4] Struck by the Garter references, Hotson —already oriented toward a 1597 date—decided to test the lines against that year's Feast: "The Garter Feast celebrated at the palace of Westminster on St. George's Day, April 23, 1597, was of exceptional splendour and solemnity. Four years had elapsed since the last election of Knights to the Order; and St. George's Day, 1597, witnessed the creation of five new Knights: Frederick Duke of Württemberg; Thomas Lord Howard de Walden; George Carey, Lord Hunsdon; Charles Blount, Lord Mountjoy; and Sir Henry Lee."[5]

Two names in this list stood out as having some possible relationship with Shakespeare and the *Merry Wives.* George Carey, Lord Hunsdon; was he not patron of Shakespeare's company and the newly appointed Lord Chamberlain? And Frederick, Duke of Württemberg; was he not that very same Count Mompelgard whom Charles Knight had suggested in 1839 as the German duke of the play? It was common knowledge in courtly circles that Frederick, ever since his 1592 visit to England, had coveted a knighthood in the Order of the Garter, and, indeed, had ex-

4 *Shakespeare versus Shallow* (Boston, 1931), pp. 111-113, 116-122.
5 *ibid.,* p. 113.

tracted a promise from Elizabeth that he would receive the honor. So persistent were his efforts to obtain it that he made himself a fit target for satire.[6]

Hotson's observation was acute. For the first time in *Merry Wives* scholarship a link had been established between the entire "duke de Jarmany" subplot, the Order of the Garter references, and the Windsor setting. This threefold association led Hotson to the conclusion that "Lord Hunsdon's servants presented the hastily prepared *Merry Wives* before the Queen at Westminster at the Feast of the Garter on April 23, 1597."[7] But before accepting this conclusion, we should examine Hotson's premises more closely since the facts on which they are based, not having been fully explored in *Shakespeare versus Shallow*, call for considerable amplification.

Hotson's hypothesis stands predicated upon three points which he himself has summarized: "I find it difficult to understand why someone before this (even without the help of the discovery concerning Justice Gardiner, which points to 1597) has not seen the Feast of the Garter on April 23, 1597, as the obvious occasion for the first production of the *Merry Wives*. All the evidence has long been either known or readily accessible: that the play was evidently written for a Garter celebration; that Frederick, the 'duke de Jarmany,' was elected to the Order in 1597; that his absence and unpopularity would make this election a juncture most apt for the satirical hits in the play; and that Lord Hunsdon, the master of Shakespeare's company, was also elected on the same occasion. Yet in the presence of all this consenting evidence for the Garter feast of 1597, the keenest and ablest commentators have variously suggested such dates as 1598, 1599, 1600, and 1601."[8]

---

[6] *ibid.*, pp. 113-115. An account of Mompelgard appears below in Chapter VII.

[7] *Shakespeare versus Shallow*, pp. 121-122.

[8] *ibid.*, p. 122n.

As Hotson points out, recognition that the play was written for a Garter celebration is not new. The stress, however, is on a celebration as distinct from an election. As will shortly be discussed, celebrations were annual affairs whereas elections were merely by-products of these annual observances called at the discretion of the sovereign. This distinction widens the range for anyone attempting to ascertain the date of the Garter function at which the play could have been performed.

The choice seems narrowed again by the other two pieces of evidence which focus directly on 1597—an election year. Collectively, the elections of the Duke of Württemberg and Lord Hunsdon do support that date, but considered separately, their validity for supporting the thesis may be open to question. When Charles Knight first identified the "duke de Jarmany" with the Duke of Württemberg, he did so in the belief that the reference had its roots in the 1592 visit of the Duke (though missing the point that Frederick was still Count Mompelgard) and therefore the reference served as evidence for composition of the play close to that date. At the other extreme, Duke Frederick was not invested with the Garter until 1603, nor installed until 1604. Thus, we must show some significance to the Duke's election in 1597 beyond what Hotson believes is a humoring of Frederick's wish by Elizabeth. Also involved is a consideration of why the German allusions in the play were recognized and undoubtedly appreciated by an Elizabethan audience.

In a similar vein, since there is no reference to Lord Hunsdon in the *Merry Wives*, we must determine why his election should be considered crucial in the dating. Was he aware that he was about to be made a Knight of the Garter? If so, did he know—as Hotson *conjectures* he did —of his forthcoming election sufficiently in advance to inform Shakespeare of the Queen's wish to see a play depicting Falstaff in love?

The pivot for answering all the questions which arise appears to lie in establishing the date of the relevant Garter celebration. Therefore, let us examine the procedures of the Order as revealed in fifteenth- through seventeenth-century documents for data which will corroborate Hotson's postulated April 1597 as the date of the Garter celebration at which the play was given its initial production. In doing so, we shall temporarily set aside other forms of evidence reflecting on the composition of the *Merry Wives*.

From its inception in the fourteenth century, the Order of the Garter has been the most distinguished chivalric order in England. Until the eighteenth century its membership consisted of the sovereign and twenty-five of the noblest knights in the kingdom. Among the twenty-five might also be found foreign rulers. Once the Order had been established, its procedures and rituals were scrupulously observed through the centuries with little alteration until comparatively modern times. The key ceremonial was the Feast of St. George which, unless prorogued by the sovereign, was celebrated annually in honor of the Order's patron saint from April 22 through April 24.[9]

Observance of the Feast of St. George commenced with the arrival of the knights during the late afternoon of the 22nd at whatever palace the sovereign was residing in at the time. Since 1567, by a decree of Elizabeth, recorded in the Garter annals (*Blue Book*, p. 54), Windsor had ceased to serve as a locale for the celebration of the Feast and was set aside exclusively for the installations of newly chosen

[9] The best histories of the Order of the Garter are Elias Ashmole, *The Institution, Laws & Ceremonies of the Most Noble Order of the Garter* (London, 1672); George Frederick Beltz, *Memorials of the Most Noble Order of the Garter* (London, 1841); Sir Nicholas Harris Nicolas, *History of the Orders of Knighthood of the British Empire. . .* (London, 1841).

knights. This action led Elias Ashmole, the antiquarian, to comment, "And to say truth, this Statute was but too strictly observed, all the remainder of her Reign; for we meet not with one Feast of St. George, held simply and peculiarly as a Feast in honor of the Order's Patron at Windesor (unless you mistakingly account any of the Feasts of Installation for those of St. George...)."[10]

Upon arrival at Court, the Knights-Companions held a meeting known as a chapter and then attended vesper services. Afterward they went to supper, a meal at which none but Knights-Companions and officials of the Order were permitted to be present. On the morrow, St. George's Day, the members participated in both sacred and secular ceremonies. These commenced about 9 a.m. with a First Morning Service which the sovereign never attended. Then, a couple of hours later, the Second Morning Service was celebrated. It was both preceded and followed by the Grand Procession of the sovereign and knights to and from the chapel. Next came a resplendent banquet of several hours' duration, after which a vesper service took place. A supper, which the sovereign seldom attended, followed. On April 24, the last day, the morning proceedings commenced with a chapter, followed by chapel services during which the Knights-Elect, if any, were invested with certain of the Garter insignia. With the completion of this service, the Feast ended. When an election had been decided upon, the nominating of the new knights took place at any of the chapters held during the course of the three days.[11]

10 Ashmole, pp. 474-475.

11 This summary of proceedings is from Ashmole, pp. 504-602 and from *The Statutes and Ordinances of the most Noble Ordre of Saint George, named the Gartier...*, *passim*. Reprints of the Statutes appear in Ashmole and in Thomas Dawson, *Memoirs of St. George The English Patron; and of the Most Noble Order of the Garter* (London, 1714). Though Ashmole was writing in 1672, he had prepared his material from the quantity of records of the Order which he had collected and which are now preserved at the Bodleian Library.

The Grand Procession on St. George's Day represented the height of pageantry of the ceremonies. Some indication of the magnificence of this procession may be gleaned from a description left by Hans Jakob Breuning von Buchenbach, the Württemberg ambassador, who was a guest at the ceremonies on St. George's Day, 1595. Breuning relates that when the Knights-Companions called to escort the Queen to the chapel:

". . . Her Majesty stepped out of the Privy Chamber, arrayed in silver cloth. On her robe were embroidered two obelisks crossed, which in lieu of a button had at the top a beautiful oriental pearl. The robe was further adorned with rare costly gems and jewels. On her head she wore a very costly royal crown. Her majesty was escorted on either side by Knights and Earls. Her train was borne by a maid-of-honour. On stepping out of the chamber Her Majesty greeted all present. Then there followed in great numbers all the countesses and other noble ladies who had awaited her in the Presence Chamber. Round her Royal Majesty were many nobles with small gilt pikes like those carried by the halberdiers of His Imperial Majesty. From the Presence Chamber the procession went into the chapel. Here were the officiating clergy who, like the papal, wore albs and dalmatics and levitical vestments of gold material. They solemnized a Mass, which in the presence of Her Royal Majesty and of the Knights of the Order lasted a good while. There was a great crush in the chapel, as many of the common people had thronged thither. Mass having been solemnized and prayers said, the Knights walked in the order described to the castle-yard, followed by Her Royal Majesty under a canopy of cloth of gold with red lining, borne on poles by four men. Her Majesty's train was this time carried by a nobleman. Then followed the womenfolk, and in this order they marched round the yard three times so that everyone could have a good view of them. Her Majesty spoke most graciously to everyone, even to those of the vulgar who fell upon their knees in homage."[12]

Following the conclusion of the Grand Procession, the Knights-Companions and sovereign (if he chose to be present) partook of the Grand Dinner.[13] The elaborateness of this ceremonial can also be judged from Breuning's remarks as he continues his account:

"When this procession was over, Her Majesty returned to the Presence Chamber, where was the most splendid array of all meats imaginable that were to be had at this time of the year and from beyond the seas. No expense was spared. There were three long tables laid and prepared. The table at the top of the room stood under a splendid canopy of cloth of gold. It was the table at which the courses are carved and served up for Her Royal Majesty, and to wit, with the same ceremoniousness in her absence as when she is present. This is done even when no one sits and dines there. At this table sat this time Mylord Cobham all alone, who at this festival had to represent the Queen. He was also served and waited upon exactly as if Her Majesty had been present in person. The Queen's guards who are always attired in red coats with black velvet facings, wear on their breasts and their backs brass roses and the name of Her Majesty. These guards served the comestibles in gilt silver plate and fell upon one knee before the table. Those Earls who handed the water before and after the meal knelt upon both knees. . . .

"The Knights sat down to the banquet at one o'clock and

12 "My . . . narration . . . from the day . . . I was . . . dispatched to England . . . till my return. . . ," *Queen Elizabeth and Some Foreigners* . . . , ed. Victor von Klarwill, trans. T. H. Nash (London, 1928), p. 378, hereafter cited under Klarwill.

13 Ashmole, pp. 500-501, 588-597, 602-612. This ceremony is also termed "Grand Feast"; since the whole holiday was named the Feast of St. George, "Grand Dinner" causes less confusion.

rose again between four and five o'clock, after two Angli-
can clergymen in the middle of the hall had made a low
obeisance and then offered up a short prayer. This had
also been done before the meal. Before leaving the room
all the Knights bowed with due reverence before Lord
Cobham. The nobles who had waited upon the Knights
all wore blue tunics of cloth and on their sleeves the arms
of their overlords. Over this coat they wore golden chains,
through which they had one arm thrust. Their dress was
otherwise partly of silver or gold cloth or at least of velvet
and silk. There are also a goodly number of burgesses who
must attend at Court in similar blue tunics at certain
times, in return for which they are exempted from all
tributes, rates and taxes and other liabilities. These, too,
dress in naught but velvets and silks, though they be but
artisans, cobblers or tailors."[14]

From the conclusion of the Breuning account it will be
noted that though technically part of the ceremonies, the
dinner ritual was celebrated with members of the public
present. This fact is attested to by other records. In 1488,
for example, Henry VII gave a memorable banquet which
the Queen, the King's mother, and the Spanish and Scottish
ambassadors attended.[15] Ashmole (p. 595) reports that at
the 1520 dinner "a little while before the second course
was carried in, Queen Katherine and her Ladies came to
the Gallery at the end of the Hall (which was nobly prepared for her) to see the honorable Services and Ceremonies there performed." The Garter records continue to
turn up references to outsiders present at the banquets.
Ashmole has found mention of foreign ambassadors, of
Pages of Honor who waited on the sovereign and of Gentle-

14 Klarwill, pp. 378-379.
15 J. Anstis, ed. *The Register of the Most Noble Order of the Garter . . .
called The Black Book* (London, 1724), II, 225-231, hereafter cited under
Anstis.

men who served the Knights-Companions. Various royal officers and servants also attended in the course of duty.[16] Among these extant lists preserved by Ashmole, one even records two physicians in attendance—a striking notion when we think of Dr. Caius hurrying to the "grand affair." And many commoners came solely in the capacity of on-lookers. A contemporary Garter account of the feast on April 23, 1632, at Whitehall, for instance, describes: ". . . a little before the Kinges second course was gone for, all the presse of the people was remoued from the rayles which compassed the Knightes Table, leaving the place spatious for the King to take them all in view. . . ; Immediately before the entrance of the second Course all the presse of the people were againe voyded to the left side of the roome" (Bodl. Ashm. MS. 1110, fols. 43v-44). Again, at the banquet of April 23, 1635, the "people removed a little before the 2d course" (Bodl. Ashm. MS. 1110, fols. 7-7v).

The significance of these accounts of the elaborate cere-monies on St. George's Day is that the ceremonies were held regardless of whether an election of new members took place. Indeed, the primary function of the annual observance was to honor St. George. So serious an obliga-tion was it for the members to attend the Feast that only the sovereign, and he by royal warrant, could excuse ab-sence.[17]

16 Ashmole, pp. 500-501, 515, 575.

17 Many records of these warrants are still extant. A few are cited by way of example. On April 20, 1598, Lord Burghley was excused "by reason of your want of health and weak estate of body yet remaining, through your late great sickness" (Historical Manuscripts Commission, *Calendar of the Manuscripts of the Most Hon. The Marquis of Salisbury Preserved at Hatfield House* (London, 1883-1940), VIII, 138-139, hereafter cited as *H.M.C. Salisbury*. Essex, though under confinement for his attempted rebellion, inquired on April 19, 1600, through his keeper Sir Richard Berkeley, "to know her Majesty's pleasure (because he is sworn to the statutes of the Order) whether he shall wear his robes on St. George's day in his dining-chamber or else privately in his bedchamber, or whether her Majesty will give him a dispensation not to wear them at all that day" (*H.M.C. Salisbury*, X, 116). Edmund, Lord Sheffield, wrote to Sir Robert

The election of new members in itself furnished no special occasion for elaborate additional ceremonies. An election, as Article xx of the *Statutes* relates, was to be called within six weeks of the death of a Knight-Companion. By Tudor times it could be held or prorogued according to the whim of the sovereign with the result that there were intervals of several years without any elections. Nominations, in contrast, regularly were made if any stalls were vacant. This nominating ceremony, known as "Taking the Scrutiny," became the business of one of the chapters held during the Grand Feast. Even then the members did not have to fill all vacant places. But regardless of whether the sovereign either unilaterally or in consultation with the Knights-Companions refrained from declaring an election, whatever nominations the Knights-Companions did make were recorded in the annals of the Order.[18] Theoretically nominations—except those of foreigners—were made according to various classes of knighthood and by a majority vote, but actual power of election was the sole prerogative of the sovereign who, upon studying the nominating ballots, "then shall chose [*sic*] of them that be named, he that shall have the moost voyces, and also he that the Soverayne shall extreme [*sic*] to be most honorable to the sayde Order, and most profitable to his Crowne and to his Realme."[19] In practice this meant that whichever nominee the mon-

---

Cecil on April 13, 1602, "Through an unlucky fall I had not long ago, by which my shoulder was thrust out of joint, I am so unable to travel that I can by no means attend her Majesty upon St. George's Feast, and must make bold to entreat you to procure for me, according to the order, her gracious license for my stay" (*H.M.C. Salisbury*, xii, 106).

18 The annals consist of a series of MS books entrusted to the custody of the Dean of Windsor, Registrar of the Order of the Garter. The volume covering the reign of Queen Elizabeth is known as the *Blue Book*, also referred to by its Latin title *Liber Coeruleus*.

19 *Statutes*, Article xx. The above details of election procedure have also been culled from the *Blue Book*, p. 42 and from Ashmole, pp. 262-279, 290-294.

arch favored, he could declare elected regardless of the results of the Scrutiny.

If a nominee were so chosen and he were present at Court, he was brought to the chapter held on the third morning of the Feast. There he went through the investiture ceremony, receiving the Garter and George (the badge of the Order), and became a Knight-Elect. Full-fledged membership came only after the installation at Windsor, held at a date set by the sovereign.[20]

The foregoing synopsis of the ceremonials of the Order of the Garter shows that the most important day of all was St. George's Day and that the "exceptional splendour and solemnity" which Hotson attributed to the election of 1597 were found equally in nonelection years. The problem, therefore, is not one of pitting election and nonelection years against each other, but rather one of determining whether any other Garter Feast during Elizabeth's reign offered conditions for the composition and performance of the *Merry Wives*.

The earliest Feast for which the play could have been written is that of 1593. That year an election was held—the only one prior to the 1597 election. The year 1593 is also the date which early critics had assigned to the *Merry Wives* on the basis of the Mompelgard visit and performance of *The Jealous Comedy*—points to be considered subsequently. The latest possible date is 1601 since the Quarto was entered in the Stationers' Register on January 18, 1602.

In 1593 at the Feast held at Whitehall, five new knights became members of the Order: Henry, Earl of Northumberland; Edward, Earl of Worcester; Thomas, Lord Burgh; Edmund, Lord Sheffield; and Sir Francis Knollys. None of these Knights-Elect had any connection with Shake-

---

[20] Ashmole, pp. 298-299, 301, 312. Throughout Elizabeth's reign an installation never coincided with the observance of the Feast of St. George.

speare or his company. More important, no "duke de Jarmany" was involved in an April 1593 Garter function. Although Count Mompelgard had visited England the previous summer, he did not become the Duke of Würt-temberg until August 8, 1593.[21] And the ceremonials them-selves seem to have been quite ordinary that year. The *Blue Book* (pp. 122-124) indicates that the Queen attended most of them and that the Grand Procession was particu-larly magnificent; beyond this, nothing extraordinary. An account of the Feast in B.M. Harl. MS. 304 (fol. 168) only confirms the details entered in the *Blue Book*. Assuming that the *Merry Wives* was written for a Garter Feast, we find, then, absolutely no links between the play and either the events of or personages involved in the 1593 celebra-tion.

The following year the Feast was held at Greenwich. The Queen, however, absented herself. She appointed Henry Carey, Lord Hunsdon to act in her stead, stating: "Wee for diuers causes vs moving may not be present in our parson at all the devine seruices, and other the Cere-monies to be done and executed this daie being th'Eve and Vigile of St George for the celebrating of our saide Order, neither at some tyme to morowe, which shalbe the daie and ffeast, or the next daie after in the Morning, That is to saie the xxij[ie] xxiij[ie] and xxiiij[ie] daies of this present moneth [of April]. . . ."[22] Though during the course of the proceedings nominations were made to fill void stalls, when the results were presented to the Queen, she de-clined to certify the election.[23] There remains another

21 B.M. Cott. MS. Vesp. F. III, fol. 177.

22 P.R.O. E 101/432/12. In absenting herself the Queen acted in ac-cordance with Article VII of the *Statutes*.

23 The *Blue Book*, pp. 125-127. In commenting on this practice of Queen Elizabeth in not making elections even when the Scrutiny had been taken, Sir Nicholas Harris Nicolas observes, "This proceeding was, however, part of Elizabeth's favourite system of government. Every vacant Honour or Office, secured the fidelity and zeal of numerous candidates; and she

factor to be considered. Plague was still rampant at the time, and a proclamation was issued on April 21, 1594, forbidding all "who either have had the sickness, or whose houses have been infected any time within these twelve months" from coming to court for the Feast of St. George.[24] Under such conditions, a special dramatic production is highly unlikely; when coupled with the Queen's absence from most of the ceremonies and the cancellation of an election, 1594 may safely be eliminated.

In 1595 at Whitehall, "At the hower accustomed the said 22th of Aprill all the Knightes assembled into the presence Chamber And the Soueraign came & proceeded to the great Closett & there held a chapter for the appointing her . . . said Lieutenant the Commission being read by the Chancellor & deliuered to the said Lord Cobham lowly vpon his knees by her Majesty."[25] The Queen then absented herself from practically all the ceremonies of this year's Feast. Even the fact that she had invited two German ambassadors—Count Philip von Solms, envoy of Maurice, Landgrave of Hesse Cassel, and Hans Jakob Breuning von Buchenbach, representative of the Duke of Württemberg —as guests at the Grand Dinner did not move her to attend. Breuning, who had come to England for the express purpose of entering another plea for Elizabeth to award the Duke the Garter, specially notes in the report of his mission that the banquet on April 23 commenced at one and ended between four and five without the Queen ever putting in an appearance.[26] Nor, as the *Blue Book* relates, did she appear during the taking of the Scrutiny that night. On the following day in concert with the Knights-Com-

preferred relying upon the hopes of many expectants, than upon the gratitude of the few who could be appointed" (*History of the Orders of Knighthood of the British Empire. . .* , I, 199).

[24] *H.M.C. Salisbury*, IV, 514.

[25] Bodl. Ashm. MS. 1109, fols. 95-95v. There is another copy among the papers of Sir William Dethick in B.M. Add. MS. 10,110, fol. 23.

[26] Klarwill, p. 379.

4. Arms, Supporters, and Badges
of Queen Elizabeth I

5. *Artist's conception of Elizabeth I in procession as Sovereign
of the Most Noble Order of the Garter*

GENTLEMEN USHERS   ROBERT HORNE,   WILLIAM DAY,   MAXIMILIAN I
                     Bishop of Winchester,   Dean of Windsor,   Emperor of Germ
                          Prelate   Registrar

          SIR THOMAS SMITH,   Usher of the   SIR GILBERT DETHICK,
              Chancellor   Black Rod   Garter King of Arms

ROBERT DUDLEY,   AMBROSE DUDLEY,   FRANCIS RUSSELL,   WILLIAM SOMER
Earl of Leicester   Earl of Warwick   Earl of Bedford   Earl of Worces

ANTHONY BROWNE,   GEORGE TALBOT,   HENRY CAREY,   SIR HENRY SYDNEY
Viscount Montague   Earl of Shrewsbury   Lord Hunsdon

der of the Garter during Elizabeth I's reign.

Depicted are the Knights-Companions who were members in 1576, as identified by their coats of arms. The portraits are not historically accurate.

HENRY III, King of France	ADOLPHUS, Duke of Holstein	FRANCIS, Duke of Montmorency	THOMAS RATCLIFFE, Earl of Sussex
'HILIP II, ng of Spain	EMANUEL PHILIBERT, Duke of Savoy	HENRY FITZ ALAN, Earl of Arundel	EDWARD FIENNES, Lord Clinton

WALTER DEVEREUX, Earl of Essex	ARTHUR GREY, Lord Grey of Wilton	HENRY HERBERT, Earl of Pembroke	CHARLES HOWARD, Lord Howard of Effingham
NRY HASTINGS, of Huntingdon	WILLIAM CECIL, Lord Burghley	HENRY STANLEY, Earl of Derby	(This last stall was vacant in 1576. The figure is the artist's addition.)

10. Arms within a Garter

THE Arms of the Sovereign
impaled with the Arms of
the Order

William Brooke,
Lord Cobham

George Carey,
Lord Hunsdon

Frederick, Duke of Württemberg

panions, the Queen decided not to elect any new knights.[27] All these facts remove 1595 from consideration.

As in the previous year, the Queen turned over the administration of the 1596 Feast, celebrated at Greenwich, to a lieutenant. The Earl of Essex now received the honor. The Queen then absented herself from most of the ceremonials although she did participate in the Grand Procession and attended the Second Service.[28] Though there had been great expectation for the election of new knights, the Queen waived filling any vacant stalls. None of the events of this year suggest 1596 as a suitable one for the composition of a play in honor of the Order. However, examination of the 1596 Scrutiny in the *Blue Book* reveals a significant foreshadowing for 1597, for both the Duke of Württemberg and George Carey appear among the nominees, and the account closes with a definite statement that the election is being prorogued to the following year.

Momentarily omitting the key year of 1597, let us continue our elimination procedure. In 1598, the Feast was kept at Whitehall. Sir Roger Wilbraham, who had come from Ireland earlier in the year, has left a vivid account of the events of this Feast.[29] The pageantry of the procession of Knights-Companions who had called upon the Queen to escort her to chapel on the eve of April 22 impressed him deeply; but notes Wilbraham, "her maiestie went not." Nor was she at the magnificent supper which followed. "The next day [St. George's Day] her maiestie went to chappell in procession under a canopie caried with 6: & that knights all." That nothing unusual occurred may be

27 *Blue Book*, pp. 127-129.
28 *Blue Book*, pp. 130-132. A copy of the commission to the Earl of Essex wherein the Queen states she cannot be present for the entire Feast may be found in a volume of unclassified manuscripts entitled *Garter: Miscellaneous* (fol. 346v) at the College of Arms and in the custody of the Garter King of Arms.
29 *The Journal of Sir Roger Wilbraham. . .*, ed. Harold Spencer Scott, The Camden Miscellany (London, 1902), X, 15-17.

surmised not only from Sir Roger's report, but also from the *Blue Book* entry for the year (pp. 136-137). The Queen absented herself from practically all ceremonies, and, as in 1594-96, delegated her authority to a lieutenant, this time to the Earl of Shrewsbury. She also decided to put off any election for that year. Between Elizabeth's limited participation and the lack of an election, the 1598 Feast is virtually precluded as an occasion for a special play performance.

Although 1599 saw the election of Robert Ratclyffe, Earl of Sussex; Henry Brooke, Lord Cobham; and Thomas, Lord Scrope to the Order, the Garter celebration in that year could not serve for an initial presentation of the *Merry Wives*. First, none of the new knights had any relationship with Shakespeare or the Chamberlain's Men. Secondly, Elizabeth was in no mood in 1599 for gaiety at St. George's Feast. Political worries had so seized the Queen that, as the *Blue Book* relates (p. 137), she ordered the Feast celebrated with a minimum of splendor. This on account of "sedition and flames of rebellion in Ireland."

The year 1600 presents a somewhat different case. The *Blue Book* (pp. 140-141) shows that the Queen had decided to prorogue the election to the following year. But lack of an election, as has been indicated earlier, is by itself insufficient reason to exclude a Garter Feast from consideration as offering the proper conditions for the first performance of the *Merry Wives*. However, in other non-election years discussed, the Queen usually refrained from attending ceremonies, and/or various events had occurred which served as a basis for eliminating specific Feasts. Now in spite of the prorogued election, the 1600 celebration at Greenwich was a gala one, for it had a guest of honor in the person of Monsieur de Chastes, the Governor of Dieppe. De Chastes had come to England to serve as proxy for Henry IV at his Garter installation at Windsor. Describing this visit to Sir Robert Sidney, Rowland Whyte wrote on April 26:

"The Feast of St. George was solemnised with more then wonted Care, in Regard of Monsieur le Chates being here, and other gallant French that accompanied him. Here were 13 Knights of the Order present, a very great Nomber of Ladies, and a great Shew of Noblemens Servants. There were no new Knights chosen at this Time. The Intertainment of the French in Court, was very great and magnificent."[30]

On the surface it may appear that the *Merry Wives* conceivably could have been part of that great and magnificent entertainment. But Monsieur de Chastes is the same French envoy who became involved in a major post horse scandal on September 4, 1596.[31] This happened just as he was leaving England—having served as assistant to M. de Bouillon, the French ambassador, in the ratification of the Treaty of Greenwich. The circumstances surrounding this incident are so startlingly similar to those described in the fourth act horse-stealing subplot of the *Merry Wives* that, as will be shown in a later chapter, I believe the affair served as the prototype for the subplot. That Shakespeare would knowingly mirror an unpleasant incident in the life of an individual of the stature of de Chastes is beyond the wildest imagination. After all, the envoy was on an Order of the Garter mission for Elizabeth's chief ally and would have been present at any play then shown at Court. Nor can last-minute ignorance of de Chastes' arrival be

30 Arthur Collins, ed. *Letters and Memorials of State . . . Faithfully transcribed from the Originals at Penshurst Place . . .* (London, 1746), II, 190. See also Historical Manuscripts Commission, *Report on the Manuscripts of Lord de L'Isle & Dudley Preserved at Penshurst Place* (London, 1925-42), II, 457, hereafter cited as *H.M.C. Penshurst*. The documents in these two collections complement each other in reproducing the Sidney Papers.

31 John Crofts, *Shakespeare and the Post Horses* (Bristol, 1937), pp. 18-21. In French records his name is spelled de Chaste.

considered a possibility, for as early as February 29, 1600, the court gossip John Chamberlain had written to Dudley Carleton, "We heare there is some great man comming out of France, in shew about the Kinges installation at Windsore, whatsoever other errand he may have in secret."[32] Thus taking all the evidence for 1600 into consideration, I find the possibility of a first performance at the Garter Feast that year a remote one.

Initial presentation at the 1601 Garter solemnities likewise appears inprobable. During the course of the Feast (celebrated at Whitehall), two new knights became members of the Order: Sir William Stanley, Earl of Derby and Thomas Cecil, Lord Burghley. Neither individual had any connection with either Shakespeare or the Chamberlain's Men.

This study of each of the Feasts from 1593 to 1601 reveals that not one of them—except that of 1597—offered the proper circumstances for the special composition and presentation of the *Merry Wives*. The claim for 1597, however, is based on much more than the results of an elimination process, as examination of the complete ceremonies for that year will further confirm.

The usual authorization for the preparation of Whitehall Palace had gone out to the Groom of the Chamber— this time to Richard Brackenbury, as the record of payment in the Audit Office rolls indicates.[33] According to the

[32] *Letters Written by John Chamberlain during the Reign of Queen Elizabeth*, ed. Sarah Williams (London, 1861), p. 68. Although Chamberlain does not identify his "great man," we can surmise that his name became known not too long after Chamberlain sent his letter and therefore in sufficient time for Shakespeare to have learned this Frenchman's identity—if Shakespeare were to have written the *Merry Wives* at this time.

[33] P.R.O. AO 1/386/35. It will be noted that the Feast was celebrated at Whitehall. In listing its locale as Westminster, Leslie Hotson was not actually in error. Apparently his manuscript source was one written by an individual who did not differentiate between the two palaces. Contemporary documents indicate that Westminster was also the over-all

report of Sir William Dethick, the Garter King of Arms, the Knights-Companions gathered in the palace about three in the afternoon on Friday, April 22. Dethick continues: "After vj of the cloke. All the knightes were assembled in Chapter. before the Soveraigne in the pryve chamber. where the Chancellor of the order did read the Comission. before the Soveraigne & her majesty gave the same to the Lo. Howarde of Effingham. to be Lieutenent. And her highness would not goe to the privat closett. Wherefore the LL. proceeded after that Chapter. & so went to the chappell."[34] There then followed the usual supper, also without the Queen's presence.

On the morrow the palace was a bustle of activity as people flocked to watch the ceremonies. Rowland Whyte wrote to Sir Robert Sidney on the 23rd itself, "I am going to wayt upon the children [the Sidney children] to Court. I borrowed yesterday of Sir Jo. Fortescu his chamber for them, that they may see the Queen in her procession."[35] And a majestic procession it was—according to the eyewitness account of the lawyer John Hawarde:

"The Royal Court being at Whitehall, great solemnity for the Order of the Garter was observed, being S. George's day. 'Firste morning Sarvice in the Chappell with solemne musike & voyces, Doc. Boole [John Bull] then playing, the Lordes of the order then presente; then Comming & retyring they make 3 congies to the seate Royalle & so departe; & some howre after they come againe before her

---

geographical heading given to royal property in that area of London (see, for example, P.R.O. LC5/182, fol. 34). To some recorders the two names could be used interchangeably for Whitehall. Thus there is an entry in Bodl. Ashm. MS. 1108, fol. 8v, "St. Georges day Thursday the 23d of Aprill 1562. Anno 4° Eliz: Regina holden at the White Hall called the Pallace of Westminster." And the Pipe Office rolls of 1597 for the Office of Works refer to "Whitehall alias the newe Pallace of Westmynster" (P.R.O. E351/3232).

34 B.M. Add. MS. 10,110, fol. 158.
35 *H.M.C. Penshurst*, II, 271.

Maiestie with all the officers of Armes, & then Commeth the Queene, three Ladyes Caryinge her trayne, which then were the Countes of warwike, the Countes of Northumberlande & the Countes of Shrewseberrye, th'erle of Bedforde carryinge the sworde before her, 6 peneyeners Carying a rich Canopye ouer her heade; & then after there seuerall Congees, there is shorte seruice, the Clergie all in there riche Copes, with the princely musike of voyces, organes, Cornettes & sackebuttes; & in like order her Maiestie goes one precession & so returns, & shee & the rest of the order offer at the highe altar, & so seruice ends, & shee departes. . . .' "[36]

Hawarde's report breaks off at this point in the proceedings, but the *Blue Book* minutes (pp. 133-135) round out the account. Here it is revealed that the Grand Dinner was served after the Second Service ended. That evening the Queen attended the chapter with the Knights-Companions and instructed them on the conditions for electing the new members. While the Scrutiny was taken she remained in private prayer. Upon receiving the results, Elizabeth ratified the elections of the Duke of Württemberg, the Lords Howard, Hunsdon, and Mountjoy, and Sir Henry Lee. Thereupon she personally invested the four English nobles with the Garter and Collar of St. George. The following morning, while the Queen worshipped in her private chapel, the Knights-Companions attended the regular morning service, at the conclusion of which the Feast ended.

From the preceding account, it appears that the 1597 Feast was solemnized with no more pageantry than any

---

[36] *Les Reportes del Cases in Camera Stellata 1593 to 1609*, ed. William Paley Baildon (n.p., 1894), pp. 74-75. There is also an official, though less embellished, account of portions of this Feast in *Collections Relating to the Order of the Garter*, MS. Saec. xvi, fol. 79. (This is a MS. volume in the custody of the Garter King of Arms at the College of Arms.) Another copy of this latter account, with minor variations, is in Bodl. Ashm. MS. 1108, fol. 63.

other. A comparison of Hawarde's report—from an election year—with that of Breuning's—from a nonelection year—only proves this. But what is striking about the 1597 festivities is that the Queen seems to have participated in the various ceremonials to a much higher degree than in any other of the years between 1593 and 1601. This observation holds particularly for her attendance at the chapter held the evening of April 23, for this was not in keeping with her practice of the four preceding years. Also, her close supervision of the election with the immediate ratification and investiture of the newly-elected knights did not follow normal procedure, which called for the ceremony of investiture during the chapter convened on the third morning of the Feast. Reasons for this unusual action can only be inferred. Perhaps the immediate investiture was motivated by the fact that Lord Hunsdon, Elizabeth's favorite cousin, was one of the newly chosen knights.

Other grounds exist for considering the 1597 election unique. For political reasons (subsequently discussed) Elizabeth had to assure the election of the Duke of Württemberg. Then there was a strong lobby instituted by the Earl of Essex for making Sir Henry Lee, who was not a peer,[37] a Knight of the Garter.

This much, however, is clear: a gala spirit pervaded the 1597 Feast. Only in 1597 was there a tie between Shakespeare's company and the Garter through the election of Lord Hunsdon, the company's patron. And the Duke of

[37] Four days after the election Rowland Whyte wrote to Sir Robert Sidney: "My Lord of Essex, as I hard, was exceading earnest with his companions, for their Voices in the Election of Sir H. Leigh, which he obtained; then had he much a doe to bring the Queen to giue her consent for hym; but soe earnest was he for hym, that he preuailed" (Collins, II, 45; also in *H.M.C. Penshurst*, II, 271). On April 30 Whyte wrote again: "I acquainted my Lady Essex, with my Buisnes [*sic*] in Court about the Horse, and desired her to speak to my Lord to take that Care of yt, as he of late did in a far greater Matter for some other of his Frends, and named Sir Henry Leigh; who meerly by my Lord Essex's Favor was receued [into the Garter]" (Collins, II, 47).

Württemberg was the sole German nobleman elected to the Order between 1579 and 1612.

With the case for 1597 established, the question arises whether it can be determined when in the proceedings a play could have been presented. Three possibilities suggest themselves: during some portion of the Feast of St. George; in the interval between the Feast and the Windsor installation (April 25-May 22); or at the installation.

Several facts gainsay a performance at the installation ceremonies. Elizabeth, since her decree of 1567 removing all Garter observances from Windsor except installations, attended no functions of the Order at the castle for the remainder of her reign. To supervise the installations, she appointed special commissioners. Yet the Quarto title page specifies that the play was presented "before her Maiestie." If this notation is accurate—and there is no reason to doubt its veracity—then the Queen could not have seen the play at a Windsor Garter celebration (i.e., an installation).

Also remote is the likelihood of an initial production at Windsor before the Queen on a non-Garter occasion, for Elizabeth appears to have shunned trips to this residence in her later years. After her visit of 1593, she did not return to the castle until August 1601, as far as extant documents reveal. By that date the *Merry Wives* assuredly had won a place for itself in the public theater. Elizabeth probably kept away from the castle with good reason, for although she liked Windsor, a contemporary reporter remarks that "the howse be colde"[38]—not the healthiest residence in England for an aging monarch. Taking into account all the evidence—internal as well as external—for the composition of the play for a Garter festivity and in response to the Queen's wish to see Falstaff in love, I find it hard to believe that the Quarto reference to presentation "before her Maiestie" can apply to aught but the first

[38] Quoted in Harwood, p. 221.

performance; and that could not have taken place at Windsor in the presence of the Queen under any circumstances.

Lest the possibility be raised of a premiere at the 1597 installation without Elizabeth in the audience, let us examine the contemporary accounts of that installation. The Queen set May 24 as the date, naming the Lord High Admiral, Charles Howard as her chief commissioner to oversee the ceremonies at Windsor and appointing as his assistants Lord Buckhurst and the Earl of Northumberland.[39] Following the usual procedure for installations[40]—which was much simpler than that for the Feast—the ceremonies extended over a day and a half. They commenced with an elaborate processional into the castle on the afternoon of the 23rd as recorded by one of the heralds present:

"First Sir Henry Lea, with his Company, came riding through the Towne from Stainesward, all his men well mounted, & in blew Coates & Badges.

"Next after him came riding the Lord Mountioy, with all his men in blew Coates, every one a plume of purple estridge feathers in their Hattes, his Gentlemen Chaines of Gold.

"Thirdly & imediately after him came my Lord Chamberlaine with a braue company of men & Gentlemen his Servantes & Reteyners, in blew coates faced with Orange coloured Taffety, & Orange coloured Feathers in their Hatts, most parte having Chaines of gold; besides a greate number of Knightes & others, that accompanied his Lordship.

"Lastly the Lord Thomas Howard came imediately after with like troupe & blew coates, faced with sad sea colour greene Taffeta, with Feathers of the same colours, & many chaines of gold, which made a goodly show, & the more,

39 *Blue Book*, p. 135.
40 Ashmole, pp. 338-366; *Blue Book*, *passim*.

for that they came all foure together in Order, & not dropping one after another, & out of Order. . . ."[41]

Upon their arrival, as another account picks up:

"All these Lordes [the four Knights-Elect and the three commissioners] with their Traynes, came very honourably to Windesor, on Monday the 23^d of May about 5 of the Clock in the Afternoone, and went to their severall Lodginges. . . . The Lord Admirall was lodged in Beauchampes Tower: And the next Morning thither came the other Lordes, all saveing Sr Henry Lee (who was spared because of his gout) and so proceeded to the Chapter house where the Commission beeing read Garter came forth and brought in the Lord Thomas Haward, where the Commissioners received him, put on his Kyrtle & hood. . . . Then the Lord Thomas Howard betweene the Lord Admirall and the Lord Buckhurst [went into the Choir]. And being come vnderneath his Stall, the Register gave him his Oathe, and so hee was brought vp and enstalled. . . .
"Then the Lordes left him there, and went back, & in like manner brought the Lord Hunsdon, then the Lord Montioy and lastly Sr Henry Lee. At Service tyme began the Offering of the Hatchmentes of 7. Knightes of the Order lately deceased . . . service beeing done, they all went downe through the body of the Church & so out, at the South dore vp into the Castle to Dynner, where honourable faire was prepared at the Charges of the 4 new Installed Knightes, whose Servantes attended and served vs (the Officers of Armes) of all things needefull."[42]

In spite of the fact that there were large numbers of attendants at the installation and that Shakespeare and his

---

[41] Bodl. Ashm. MS. 1112, fol. 16v; another copy in B.M. Stowe MS. 595, fol. 45v; also quoted in *Shakespeare versus Shallow*, pp. 118-119.

[42] Bodl. Ashm. MS. 1108, fols. 74v-75. This is a transcription, with minor modifications, from what I believe is the original manuscript in Coll. Arm. *Collections Relating to the Order of the Garter*, MS. Saec. XVI, fols. 87-88.

company may have been among the three hundred in Lord Hunsdon's retinue,[43] the above accounts do not offer evidence suggesting a performance of the play at Windsor. In fact, they do just the opposite, for aside from confirming the absence of the Queen, they establish that only seven Knights of the Garter attended the Windsor ceremonies whereas the full complement of able members was at Whitehall.

With the installation eliminated, the month preceding it appears a possible period for the first performance. The Feast ended on April 24. On May 7, the Court moved to Greenwich. However, not all the Garter knights went with it. Lord Burgh, who had delayed his departure on a government mission to Ireland solely to attend the Garter Feast, commenced his trip on the 25th of April.[44] In this action lies the clue to ruling out a performance in this interim period. Since the Knights-Companions were not compelled to remain at Court once the Feast officially closed on April 24,[45] it seems improbable that a performance of a Garter play would have been planned when no guarantee of a Garter audience existed.[46]

We are thrown back, therefore, to the Feast proper, where further delimiting is possible. The 1597 Feast, as usual, did not get under way until late afternoon on April 22. A supper for the Knights-Companions and Officers of the Order—the only ones permitted to attend that function—followed the vesper service. Since the Queen was not present at the service, and since she did not customarily sup with the Knights that first evening, no opportunity for performing a play before Her Majesty presents it-

---

43 See below, p. 62 n.1.

44 *H.M.C. Penshurst*, II, 265 (Letter of April 13).

45 Ashmole, p. 632.

46 Even the Gentlemen of the Chapel were at liberty to leave the Court for "a weeke after St. George" (*The Old Cheque-Book . . . of the Chapel Royal . . .* [Westminster], 1872, p. 73).

self. Furthermore, the supper probably took place in the chambers of Lord Howard of Effingham.[47] Private chambers, an audience of ten to thirteen,[48] and the absence of the Queen do not suggest the proper surroundings for a special performance. Similarly, April 24 is not favorable, with the Feast ending about midday and the morning entirely devoted to ceremonials. This leaves St. George's Day itself—a most appropriate occasion when we remember the long tradition of pageants and processions held throughout England on that particular day.

Every year on St. George's Day, unless the Feast were prorogued, after the Second Morning Service the Grand Dinner was held. Consisting of two "courses," each featuring about eighteen distinct dishes, this dinner—as Breuning's account mentioned—lasted until four or five in the afternoon. Ashmole (p. 597) even has a record "so long did the services of the Dinner hold, when the Feast of St. George was celebrated at Windsor, the 14. of September, an. 15. Jac. R. that the Knights-Companions proceeded by Torch light to the Chapter-house." And at these banquets a great many people having no connection with the Order were in attendance.

The *Blue Book*, unfortunately, has few details of the Grand Dinners. Neither does the earlier *Black Book*. This paucity of references leads one to believe that the chroniclers of the registers were instructed to enter only the business sessions of the Order. When embellishing details creep into their accounts, these details imply that entertainment at a Grand Dinner would not have been taken as anything

[47] Bodl. Ashm. MS. 1109, fols. 35v-36v. An entry in this MS, which records the lieutenants at the various feasts, states that in 1597 Lord Howard of Effingham "made a dinner in his Chamber." This reference presumably can apply only to the first night supper.

[48] The *Blue Book*, p. 133, records ten Knights-Companions at the Feast. There were only three officers at the time: Sir William Dethick, Garter King of Arms; Sir Edward Dyer, Chancellor; Robert Bennet, Registrar. The posts of Usher of the Black Rod and Prelate were vacant.

exceptional, indeed hardly worth a passing word. Indicative of this attitude is a *Black Book* entry for May 27, 1535, about "keeping the Feast in a most glorious manner"[49] with no further elaboration.

Precedents, however, do appear for some form of entertainment at the banquets. The *Black Book* records an event during the reign of Henry VII at the Garter celebration of 1488:[50] "Divine Service being over by Noon, the Sovereign sate down to an Entertainment in a splendid Manner." After describing the seating arrangements and identifying the distinguished guests present, the author of the account continues:

"Whilst these Things were done with all kind of Magnificence, as well in the Household Stuff and Ornaments, as in the Meat and Services, the King retired from the Entertainment to his Bed-Chamber, that after resting, he might return to dispatch what remained with the like Decorum.

"I say nothing of the Songs, the Sonnets, and Rhimes, published every where in Praise of the King, and on the Happiness of the Kingdom under such a Prince; I say nothing of many other Things, the Sight whereof is more affecting than the Narration; this only be assured of that no kind of Magnificence was here omitted."

Would that our chronicler had not been so reticent! But a transcript of the events of this feast remains which concludes with a series of verses honoring King Henry and the Order of the Garter. Their heading reads, "These verses were presented at this present ffeast of St. George at Windsor."[51]

---

49 Anstis, II, 393. Another example is an account found in B.M. Harl. MS. 304 for the Grand Dinner held on April 23, 1593: "Dinner also was served accordingly [and] at the 2 Course Garter proclaimed the Stiles etc." (fol. 168).

50 Anstis, II, 229-230.

51 See Bodl. Ashm. MS. 1131, fols. 68v-71.

A further precedent is found during the reign of Henry VIII at the prorogued banquet held on May 29, 1520. Here "the trompetts blew for the most part of the dyner tyme, except at such tyme as oder mynstrells of England and of Spayne did playe."[52]

Here then is the Grand Dinner spreading over four to five hours during a primarily secular ceremonial of the Garter on St. George's Day. At it is a ready-made audience for witnessing entertainment. Here also can be found specially composed poetry, minstrelsy, ceremonial music—all calculated to make that dinner a truly splendid affair. May not the Grand Dinner therefore offer a possible opportunity for presenting a play? True, no record of a dramatic production in the course of these Grand Dinner entertainments exists, but neither had there been a Shakespeare at the Court whose patron was about to become a Knight of the Garter.

Still another—and more appropriate—occasion on St. George's Day presents itself for consideration. At the conclusion of the Grand Dinner, the Knights-Companions proceeded to the chapter house for a meeting and/or to the chapel for evensong. Afterward they attended a supper, which, coming upon their banquet, would have been less elaborate. Once they had supped, the Garter knights and other nobles attendant at Court may have sought some light entertainment before retiring as a welcome change from the rituals and processions which had filled their day.

That such a suggestion is not pure conjecture can be proved by noting the manner in which the Earl of Leicester observed the 1586 Feast of St. George in Utrecht when his duties as "Lieutenant and Governor-General of her Majestie's forces in the Low Countries of the United Provinces" prevented him from attending the celebration in

[52] Anstis, II, iv, Appendix ii. Reprinted from a manuscript described as M. 17 in Offic. Arms.

England.[53] On April 23, bound by the statutes of the Order, Leicester carried out the usual ceremonials in Utrecht. First he performed the religious rites in the cathedral, and then gave a magnificent feast at his residence. Even here he followed protocol by having a proclamation of stiles and the crying of "Largesse" three times before the second "course." Then: ". . . the feast ended, and tables voyded, there was dauncing, vaulting and tumbling . . . and thus they passed the time till evensong, and then departed. At supper being all assembled againe, great was the feast, and plentifull the cheere; and after supper beganne the barriers betweene challengers and defendants, men of armes, wherein the Earle of Essex behaved himself so valiantly, that he gave all men great hope of his noble prowesse in armes.

"The barriers done, and eyther part retyred with equall prayse (though not with equall blowes), there was a most sumptuous banquet prepared of sugar meats for the men of armes and the ladies; which banquet being furnished, my Lord, wishing them all good rest, tooke his leave; and so this honorable feast broke up about twelve of the clock at midnight."

Since the Garter rituals were so solemnly observed, it seems hardly likely that this merrymaking would have taken place unless the Earl of Leicester knew he would not be going counter to established practice. In 1597 a similar evening of pleasantry surely is in order. Only instead of jousting, a new play—specially written—would serve as the evening's diversion.

Why should a new play be chosen to round out this particular evening? Why should Shakespeare have been the one to write it? The answer comes in turning back the clock a few hours—before the start of the supper. Lord Hunsdon, the patron of Shakespeare's company, had just

53 Nichols, II, 455-457.

been declared a newly elected Knight of the Garter. Moreover, he had been invested by the Queen's very own hands
with the Collar of St. George and the splendid Garter of the
Order. Thus he crowns the serious events of the day by
making this gracious gesture to his sovereign. He presents
the Lord Chamberlain's Men in a play which he commissioned Shakespeare to write especially for the occasion, a
play which both honors the Order and serves to fulfill a
wish of Elizabeth's—long recorded in the stage tradition
associated with the *Merry Wives*—that of seeing Falstaff in
love.

The 1597 Feast, therefore, appears to be the only Garter
celebration between 1593 and 1602 at which the *Merry
Wives* could have been presented. Of the three days of the
Feast, St. George's Day—April 23—alone provided an opportunity for offering the play. On the 23rd a night performance, following the supper rather than one in the
afternoon during the Grand Dinner, suggests itself for the
première. Such a time follows the usual schedule for plays
presented at Court, for, as E. K. Chambers indicates (*The
Elizabethan Stage*, I, 225), court plays customarily commenced about 10 p.m. and ended around 1 a.m. This evening presentation, moreover, offered the most appropriate
occasion for scheduling the initial production of the *Merry
Wives*, for at the time of the Grand Dinner Lord Hunsdon
was not yet a member of the Order. However, by evening
not only had he been chosen, but he had been duly invested as a Knight-Elect. His immediate investiture on St.
George's Day instead of on the accustomed last morning
of the Feast may have been arranged expressly to permit
him to appear as a Garter knight among the courtiers at
the performance of the play. Hence, on the evening of
April 23, 1597, Hunsdon, bedecked with his new insignia,
could sit among the splendid assemblage as he watched his
own troupe perform for the first time *The Merry Wives
of Windsor*.

CHAPTER III + LORD HUNSDON AND HIS NEW HONORS

ASIDE from Shakespeare, the man who probably had the most to do with the birth of the *Merry Wives* was George Carey, the second Lord Hunsdon. His inspiration in commissioning Shakespeare to write a play especially for presentation at the Feast of St. George, and on the theme of Falstaff in love, stemmed from a favorable turn of the wheel of fortune for Hunsdon in early 1597.

At the time of that turn, which carried him to the pinnacle of his career as a servant to Elizabeth, Hunsdon had been patron of Shakespeare's company for about eight months. He had taken over sponsorship of the troupe after the death of his father—Henry Carey, the first Lord Hunsdon—on July 23, 1596. Thus, as patron of Lord Hunsdon's Men—as the troupe was renamed—Hunsdon became one of the few individuals surrounding the Queen who had among his retainers an outstanding company of players and who was at the same time a Court intimate. In this dual capacity Hunsdon was in a position to act on any desires the Queen may have expressed about plays she might like to see, assuming the normal request to the Master of the Revels had not been made.

Hunsdon needed no ulterior motives for any actions he might take in attempting to gratify Elizabeth's wishes, for he was one of her devoted followers. The close relationship between both the Lords Hunsdon and Elizabeth is well known. Henry Carey was a first cousin to the Queen on her mother's side. Through this family tie the Careys held a favored position with Elizabeth, one based as much on mutual affection as on capabilities. Both Henry and George, his eldest son, so distinguished themselves in their sovereign's service that the Queen received each into the Order of the Garter. And each climaxed his career by holding the post of Lord Chamberlain.

Hunsdon's position at Court was indeed unique. His multiple roles as favorite cousin, important functionary, and master of a leading dramatic company take on significance when we consider the stage tradition which has clung so tenaciously to the *Merry Wives*: that the Queen ordered Shakespeare to write the play. If the tradition is reliable, Hunsdon would have been the logical person at Court to have served as intermediary in carrying out the Queen's command.

The tradition is first recorded in John Dennis' dedicatory epistle to his play *The Comical Gallant* (London, 1702), a reworking of the *Merry Wives*. Dennis writes: "I knew very well, that it had pleas'd one of the greatest queens that ever was in the world. . . . This Comedy was written at her Command, and by her direction, and she was so eager to see it Acted, that she commanded it to be finished in fourteen days; and was afterwards, as Tradition tells us, very well pleas'd at the Representation."

A second reference appears in the prologue of the play:

> But Shakespear's Play in fourteen days was writ,
> And in that space to make all just and fit
> Was an attempt surpassing human Wit.
> Yet our great Shakespear's matchless Muse was
>   such,
> None e'er in so small time perform'd so much;
>   • • •
> His haste some errors caus'd, and some neglect,
> Which we with care have labour'd to correct,
> Then since to please we have try'd our little art,
> We hope you'll pardon ours for Shakespear's part.

Dennis makes a third mention of the time factor—even

reducing it—in one of his Letters: "Nay, the poor mistaken Queen her self encouraged Play-Houses to that degree, that she not only commanded *Shakespear* to write the Comedy of the *Merry Wives*, and to write it in ten Days Time; so eager was she for the wicked Diversion . . . ."[1]

Nicholas Rowe, in his 1709 edition of Shakespeare, supplies a possible reason for Queen Elizabeth's command: "She was so well pleas'd with that admirable character of *Falstaff*, in the two Parts of *Henry* the Fourth, that she commanded him to continue it for one Play more, and to show him in Love. This is said to be the Occasion of his Writing *The Merry Wives* of Windsor. How well she was obey'd, the Play it self is an admirable Proof."[2]

Charles Gildon, writing the following year in his "Remarks on the Plays of Shakespear," adds further to the legend: "The *Fairys* in the fifth Act make a Handsome Complement to the Queen, in her Palace of *Windsor*, who had oblig'd [Shakespeare] to write a Play of Sir *John Falstaff* in Love, and which I am very well assured he perform'd in a Fortnight; a prodigious Thing, when all is so well contriv'd, and carry'd on without the least Confusion."[3]

In evaluating this tradition, as with all such traditions, one must exercise caution. Skepticism is healthy; but if there are strong facts lending credence to the legend and no definite contradictory evidence, the tradition deserves to be respected. This *Merry Wives* tradition attaches itself to the play eighty-six years after Shakespeare's death. How Dennis got it is unknown. Malone has conjectured that it came "from Dryden, who from his intimacy with Sir Wil-

[1] *Original Letters, Familiar, Moral and Critical* (London, 1721), II, 232.
[2] *The Works of Mr. William Shakespear* (London, 1709), I, viii-ix.
[3] In *The Works of Mr. William Shakespear* (London, 1710), VII, 291. This is a supplementary volume to the Rowe edition.

liam Davenant had an opportunity of learning many particulars concerning our author."[4] Where Rowe received his inspiration for adding the information that the Queen wanted to see Falstaff in love lies equally in the realm of conjecture. But a factual basis for the tradition can be postulated.

A closer look indicates that this tradition revolves around three points: the Queen requested the play to be especially written; only a short time was allowed for its composition; and the central theme was to show Falstaff in love. Let us relate these points directly to events at Court.

When did the Queen have the opportunity to become so enamoured with the character of Falstaff? During the 1596-97 Christmas season Lord Hunsdon's Men performed at Court on December 26 and 27, 1596; January 1 and 6, 1597; and February 6 and 8. E. K. Chambers believes that *Henry IV* (possibly both parts) may have been performed at this time.[5] This suggestion is feasible and provides a foundation for the Falstaffian element in the tradition.

But as for the Queen's giving a direct command to Shakespeare to write the comedy, and in fourteen days, here in all probability the intervening years have distorted what actually occurred. Admittedly journeying further into the land of hypothesis, I believe this is what may have happened. The Queen expressed her delight with Falstaff, casually mentioning that she would like to see the rogue in love. Either directly or indirectly Lord Hunsdon was aware of his sovereign's reaction to this new character in Shakespeare's gallery, and he promised himself to do something about it when an occasion should arise—much in the spirit of "your wish is my command."

[4] *The Plays and Poems of William Shakspeare*, I, Part ii, 190n.

[5] *The Elizabethan Stage*, II, 195-196. As will be shown in Chapter ix, there is ample evidence for dating the *Henry IV* plays earlier than now usually accepted.

A few months after the conclusion of the Christmas-Shrovetide play season the occasion presented itself. Within the same week, on April 17 and 23 respectively, Lord Hunsdon was appointed Lord Chamberlain and elected a Knight of the Garter. Did Hunsdon know sufficiently in advance that he was to receive these new honors to request Shakespeare to complete a play in time for the Garter celebration?

There cannot be the slightest doubt that Lord Hunsdon was fully aware that he would be named a Knight of the Garter at the April election, especially now that his father had died. Though few fathers and sons have been members of the Order at the same time, membership tends to follow family lines.[6] But even before he had succeeded his father to the title, George Carey had been a candidate for the Order. Examination of the Scrutiny lists in the *Blue Book* reveals that George had been nominated every year from 1593 to 1596, with the 1594 nomination coming from his own father. Whatever the reason for his not being chosen in 1593 and with prorogued elections over the next three years, by 1597 the way was open to him. His father had died the previous July; and the Queen, at the conclusion of the 1596 ceremonies, had committed herself to a 1597 election. Thus July 23, 1596, the date of his father's death, presents the earliest possible occasion for the new Lord Hunsdon to have prepared himself for his long-awaited Garter election.

That George Carey would have been kept in ignorance of his candidacy for membership in the Order throughout the previous four years would be counter to the workings of Elizabeth's Court where even the most confidential in-

---

6 See Bodl. Ashm. MS. 1134, fol. 46 for a list of fathers and sons who have been Knights of the Garter at the same time. For the best listing of the Knights of the Garter, see George F. Beltz, *Memorials of the Most Noble Order of the Garter* and Sir Nicholas Harris Nicolas, *History of the Orders of Knighthood. . .* , Vol. I.

formation seems to have rapidly become common knowledge.[7] Not even the Order of the Garter was immune to leaks about chapter proceedings. Nor was it free from external meddling. Rumors circulated openly in court circles about possible candidates. As the election of Sir Henry Lee demonstrates, much lobbying—internal as well as external—took place. Furthermore, candidates sometimes knew in advance that their elections were assured.

Thus Thomas, Lord Burghley, writing on April 22, 1601, the day before his actual election, informed his halfbrother Sir Robert Cecil, "I will now think of my instalment at Windsor. . . ."[8] And Edward Truxton made a plea in a letter to Sir Robert Cecil on April 24, 1602, for the latter to intercede with the Queen for the Earl of Hannow: "I know that with little encouragement he will tender his services to her Majesty, and become a humble suitor for the honourable order of the Garter. He will be guided herein by your Honour's counsel and direction."[9] Also, George Gilpin wrote to Sir Robert Sidney from The Hague on April 12, 1599 (almost two weeks before the election): "[I] long much till I shall hear whither any officers or counsellors shalbee made, *as was looked for*, which St. Georges daye maie chance to bring with it."[10]

Similarly speculating about a coming election, Anthony Bacon, brother of Francis and intimate of the Earl of

[7] To forestall any leak of the Scrutiny results for 1598, Elizabeth forbade the entry and all discussion of the Scrutiny for that year (*Blue Book*, p. 137). The explanation for this step can be deduced through a note from Lord Henry Howard to the Earl of Southampton on April 27, 1599: "The Queen excluded my Lord Keeper [Sir Thomas Egerton] from nomination in this last choice of knights, and though she named him not, yet gave cause to some to conceive that his being named at the election before was the cause why she would not suffer any enrolment of the scrutiny. Keep this to yourself, I beseech you, or I might be made a reporter of his disgrace whom for his virtue and his kind love to my dear Lord, I love and honour" (*H.M.C. Salisbury*, IX, 439).

[8] *H.M.C. Salisbury*, XI, 174-175.

[9] *H.M.C. Salisbury*, XII, 121.

[10] *H.M.C. Penshurst*, II, 361. Italics mine.

Essex, comments on Hunsdon's candidacy in a letter dated
April 16, 1597, to Sir Thomas Chaloner. Bacon notes, "It
was certainly thought, that the lord Hunsdon would be
lord chamberlain, and the earl of Sussex, the lord Montjoy,
the lord Thomas Howard, and the lord Hunsdon honour'd
with the garter."[11] This letter leaves no doubt that at least
one full week before his election Hunsdon knew he would
become a Knight of the Garter.

Couple this fact with those already established—the continued candidacy of Hunsdon from 1593 through 1596; the
statement at the end of the *Blue Book* entry for April 1596
that an election was definitely set for 1597; the death of
Hunsdon's father in July 1596; the fact that Hunsdon was
now patron of Shakespeare's company—and the conclusion
points in one direction. Lord Hunsdon had sufficient time
to request Shakespeare to compose a play for performance
on St. George's Day, 1597. That Shakespeare knew his
patron was to become a Knight of the Garter at this time,
even if Hunsdon had not put it to him so bluntly, is fairly
certain.

But was membership in the Order sufficient in itself to
motivate Hunsdon to request a new play, or might he have
had additional reason to make a gala celebration the evening of April 23? A second glance at Anthony Bacon's
letter of April 16 draws our attention to the line "it was
certainly thought, that the lord Hunsdon would be lord
chamberlain. . . ." And the very next day, April 17, Hunsdon received the white staff of office of the Lord Chamberlain.[12] Again the problem of timing arises—the question

---

[11] Thomas Birch, ed. *Memoirs of the Reign of Queen Elizabeth* . . .
(London, 1754), II, 331. In all probability the fact that the Earl of Sussex
was not elected in 1597 can be attributed to the lobbying that Essex under-
took for the election of Sir Henry Lee. Sussex became a Knight of the
Garter at the next election in 1599.

[12] John Roche Dasent, ed. *Acts of the Privy Council of England* (London, 1890-1907), XXVII, 50. E. K. Chambers inadvertently lists this date as
March 17, 1597, in *William Shakespeare*. . . , I, 64. The error was also made

of whether Hunsdon had sufficient foreknowledge of his new appointment to have alerted Shakespeare to prepare a play for the Court.

The earliest hint that I have uncovered that Lord Hunsdon was a candidate for the Lord Chamberlain's post is found in a communication from Rowland Whyte to Sir Robert Sidney, dated February 18, 1597—ten days after the close of the Shrovetide play season. Whyte informs Sidney, "My Lord Chamberlain [Lord Cobham] is sayd to be very ill, as sone as I can have any meanes to come to 1000 [Earl of Essex] presence I will desire hym to have you in remembrance for the cinq portes. My Lord of Hunsdon is thought shalbe Lord Chamberlain by his death, or by resignation if he live, for his body is to weake to brave the burden of the place."[13] Since Sidney had a particular desire for the Cinq Ports wardenship, Whyte sent him constant dispatches for several weeks on the declining state of Cobham's health.[14] Then on March 6 Whyte wrote: "About Midnight my Lord Chamberlain died. . . . The Court is now full of, who shall haue this and that Office; but the most Voices say, that Mr. Harry Brooke shall haue Eltam, and the Cinq Portes, by Reason of the Fauor the Queen bears hym. . . . My Lord of Hunsdon is named for the Lord Chamberlanship, and Lord Liftenant of Kent. . . ."[15] On the twelfth Whyte noted, "Report goes that Lord Hunsdon shall be Lord Chamberlain. . . ."[16] Whyte continued to keep Sidney posted on the maneuvering for offices, again commenting on April 4, "My Lord Hunsdon is like to be Lord Chamberlain. . . ."[17] By April 13, Whyte

---

in the DNB and has been perpetuated in countless editorial notes to Shakespeare's plays.

[13] *H.M.C. Penshurst*, II, 234.

[14] In addition to letters cited, see those dated February 21, 22, 25, 27, 28, March 1, 2, and 4 in Collins, II, 17-25.

[15] Collins, II, 25.

[16] *H.M.C. Penshurst*, II, 248.   [17] *H.M.C. Penshurst*, II, 263.

had noticed a distinct step forward, for now "My Lord of Hunsdon waites, and doeth all Things appartaining to the Place, but hath not yet the whyte staff."[18] Then came the appointment four days later with Whyte's comments in his letter of April 18-19, "On Sunday [April 17] 'Lord Hunsdon had the whyte staffe given hym and therby Lord Chamberlain;' he was sworn councillor and signed many letters that day. No other Aduancement or Honor was giuen to any."[19]

What these references indicate is that as early as February 18, 1597, it was bruited about in court circles that Hunsdon was next in line for the office of Lord Chamberlain. Until the death of Lord Cobham such talk must be considered only as idle gossip. However, as of March 6, with the office now vacant, the rumors become more persistent. Finally, one week before Hunsdon becomes a Knight of the Garter their truth becomes apparent. Thus as with the time element in the sequence of events leading up to his receipt of the Garter, so it is that in his assuming the office of Lord Chamberlain, Hunsdon had at least three weeks' foreknowledge of his appointment.

The two crowning achievements in Lord Hunsdon's life occurred on April 17 and 23, 1597. To become a Knight of the Garter and Lord Chamberlain in the same week is sufficient reason for any man to desire to commemorate these events in a manner that would show appreciation to his sovereign. When that individual happens to be a favorite relative of the Queen as well as master of a leading dramatic company of the day, he would be especially sensitive to the Queen's desire to see a play showing Falstaff in love. Accordingly, I postulate that Hunsdon commissioned Shakespeare to write a work centered on that theme,

---

[18] Collins, II, 38; also in *H.M.C. Penshurst*, II, 265.
[19] This entry has been collated from *H.M.C. Penshurst*, II, 267 and Collins, II, 41.

and asked him to have the script ready for presentation on St. George's Day.

Since neither of Hunsdon's new honors came as a surprise to him, he had at least three or four weeks to set this entire project in motion. John Dennis' statement, then, that the *Merry Wives* was "finished in fourteen days" does have a factual basis behind it. However, I believe that the fourteen days must be put within the larger framework of the three to four weeks. This means that the time was allotted in such a way that fourteen days went into the period of composition and that the balance was utilized for preparing the script for production and for rehearsing in accordance with the scheme described in the following chapter.

CHAPTER IV ✦ PREPARING THE FIRST PERFORMANCE

P RESENTATION of entertainments at Court fell within the province of the Lord Chamberlain's department of the royal household. With Lord Hunsdon supervising that department as well as serving as patron to Shakespeare's company, the way was clear for a liaison between the Court and the Lord Chamberlain's Men to bring off the performance of the *Merry Wives* the night of April 23. However, such a special presentation was the exception rather than the rule in Elizabeth's court. Coming after the close of the court play season (which ran from Christmas through Shrovetide during Elizabeth's reign), it would have been in the category of an "extraordinary" production. The Lord Chamberlain, therefore, would have been faced with a variety of problems not encountered during the normal playing period. Meeting the costs of production; selecting a suitable place for presenting the play when so much of Whitehall Palace was taken up with the special events of the Garter celebration; assigning musicians and singing boys to supplement the acting company; even licensing the comedy—these are some of the administrative tasks which Lord Hunsdon had to face.

Key among these tasks would have been that of paying the costs of the production. To what extent Hunsdon personally would have been liable for such expenditures we have no way of knowing. But we can be sure that the issue was not inconsequential at this time, for Hunsdon knew that his Garter induction would make heavy financial demands upon him. Not only did he have to outfit his train for the installation cavalcade and pay for their quartering in Windsor,[1] but he had to supply various ceremonial items for the installation rites.[2] In addition, each Knight-Elect, upon arrival at Windsor, had to pay his initiation fee according to his "quality and degree" and to furnish gratuities to the castle servants.

Faced with all these expenditures between April 23 and May 24, Hunsdon, through his mother, borrowed a large sum of money on May 11, 1597.[3] In the absence of any Household Books of Lord Hunsdon, we can only surmise

[1] And a large train it was, as Rowland Whyte informs Sir Robert Sidney by letter on May 14: "Vpon Monday come senight the 4 new Knights are to be installed; it was agreed vpon between them selves that they wold have but 50 Men a Piece, but now I heare that my Lord Chamberlain will have 300, and Sir Hen. Leigh 200; the other two hold their first Purpose, but they shalbe all Gentlemen" (Collins, II, 51). Also quoted in *Shakespeare versus Shallow*, p. 117.

[2] Each knight had to provide a banner, a sword, a helmet, a crest, a stall plate, lodging escutcheons, and certain apparel and regalia. The cost of the dinner following the installation was also assumed by the new members. Information on the installation expenditures has been obtained from Ashmole, pp. 312-337, 364-365, 367-368, 455-457, 462; B.M. Cott. MS. Titus C. x, fol. 15; B.M. Add. MS. 10,110, fols. 41, 88, 103v, 124v, 142v, 150v; for a comparison with late seventeenth century expenses, see B.M. Harl. MS. 1776, fols. 2v-4.

[3] *Shakespeare versus Shallow*, p. 117. I am indebted to Dr. Hotson for his assistance in guiding me to the appropriate documents which are among the Entries of Recognizance (P.R.O. LC4/193, fol. 242). Though the records note that Lord Hunsdon borrowed £2,000 from Anne Carey who had borrowed the same amount from George Bland and R. Lewishe, the sum actually loaned was about a thousand. Dr. Gerald Aylmer of the Department of History of the University of Manchester has explained to me that the Recognizance records give shortened form entries for borrowers only. The full form which the lender usually used had wording to the effect that if the loan were not repaid by a certain date, an additional stipulated sum—ordinarily about double—would be due. The Recognizance entries, however, state the full penalty total.

Oddly enough, the Entries of Recognizance also record a loan to another of the newly created Knights of the Garter. On June 12, 1597, three weeks after the installation, Lord Thomas Howard of Walden borrowed £6,000 (in actuality, about 3,000). (P.R.O. LC4/193, fol. 259.) It is possible (1) that the installation expenses were greater than Lord Howard anticipated; (2) that his loan—so much larger than Hunsdon's—had nothing to do with installation expenditures and was taken out at this time by pure coincidence; (3) that Hunsdon's loan also had nothing to do with the Garter ceremonies—though the lesser amount and the necessity of providing for the three hundred retainers incline one towards the opposite point of view.

that the purpose of the loan was to meet these expenses.[4] And we have no way of telling whether he used any of this borrowed money to defray the production costs of the *Merry Wives*—if the play was produced for the occasion and in the manner I have postulated throughout this treatise.

Whatever Hunsdon's financial state, there was one means by which he could have reduced his outlays for the *Merry Wives* to a bare minimum. Since the play was to be presented during the Feast of St. George, Hunsdon could have charged his production costs against the Feast expenditures. Ever since Elizabeth had removed the celebration of the Grand Feast from Windsor in 1567, she had assumed all its costs,[5] authorizing payment through the Treasurer of the Chamber. The Queen never drew on public funds in the Exchequer to meet any sort of Garter expenses. If, then, Hunsdon decided to take advantage of the facilities already paid for in the course of preparing the Feast, we may have the reason why the Treasury Accounts—which provide a fairly accurate index of all plays presented at Court during Elizabeth's reign—give no hint of a performance of the *Merry Wives* on St. George's Day, 1597. For, as E. K. Chambers has pointed out, "It is possible that a few [plays] may have escaped notice [in the Treasury rolls] owing to the absence of a 'reward,' or conceivably the charge of a reward to funds other than those covered by the very complete accounts of the Treasurer of the Chamber."[6]

---

4 Anthony Bacon, commenting on Hunsdon's actions at this time, wrote Dr. Hawkyns on May 7 that "the lord Hunsdon, lord chamberlain and knight of the garter, flaunted it gallantly. . . ." (Birch, *Memoirs of the Reign of Queen Elizabeth*. . . , II, 343.) Hotson erroneously attributes this letter to Francis Bacon in *Shakespeare versus Shallow*, p. 117.

5 Ashmole, p. 531. See also *Statutes of the Order*, Article VII. In Henry VIII's reign the sovereign did not meet these expenses on all occasions although the trend had set in. Aside from the items which the royal purse paid for such as the Feast and the robes for new knights, other costs were borne by the Knights of the Garter directly.

6 *The Elizabethan Stage*, I, 214.

With the premises (1) that certain production costs for this "extraordinary" performance of the *Merry Wives* could have been absorbed as part of the outlays for the 1597 Feast of St. George, and (2) that as Lord Chamberlain of the royal household Hunsdon could have directed the household retainers to assist in presenting the play, let us make a conjectural study of how the *Merry Wives* was prepared for its initial production.

According to the Cofferer of the Household Accounts, the 1597 Feast of St. George cost £256-11-6-½.[7] That this figure differs little from the cost for other Feasts should not startle us,[8] for it demonstrates that an "extraordinary" performance of a play on St. George's Day called for no additional outlay of money for purely administrative expenses or for setting up a hall for presentation. The reason for this is that the entire staff of the royal household were on duty during the three days of the Feast and received special compensation for their services. The Heralds at Arms, for example, gained an extra one hundred shillings for St. George's Day as "their accustomed largesses at principall feastes" (P.R.O. AO 1/386/35).

In this manner Richard Brakenbury, the Usher of the Chamber, and his staff received a fee of 118 shillings "for makinge readie an alteracon at whitehall of the Chambers & Chappell againste Easter and St Georges daye for her maiestie two severall tymes by the space of vj dayes menses martij et Aprilis anno xxxix^{no} Regini Regine Elizabeth . . . ."[9]

Thus it was easy for Hunsdon to have requested Brakenbury to prepare a hall for the playing of the *Merry Wives*.

[7] P.R.O. E351/1795, fol. 58v.

[8] For 1593 the cost was £261-7-4-½; for 1594, £298-6-4; for 1595, £319-0-3-½; for 1596, £253-19-8-½. Commencing with 1598, a new accounting system was instituted so that a comparison in figures serves no purpose.

[9] P.R.O. AO 1/386/35. Entries of a similar nature occur annually in the Treasury Accounts, with payment authorized by the Lord Chamberlain.

And since the Usher and Grooms of the Chamber were expert at "wayting and attending at the plaies" during the regular court dramatic season, they could be trusted to do a competent job. Which chamber at Whitehall they may have gotten ready is impossible to specify. Extant records give absolutely no clues.[10]

The responsibilities of the Usher of the Chamber would have proceeded no further than preparing the appropriate chamber and serving at the performance. Hunsdon had to rely on the Revels Office to provide the necessary costumes, properties, and items of scenery as that Office had done for the "extraordinary" performance of a masque in 1572 before the Queen and the Duke of Montmorency, who had come to England to be installed in the Order of the Garter.[11] The Revels staff had performed a similar task in

10 We know from Breuning's account of the 1595 Feast that the Grand Dinner that year was served in the Presence Chamber. And Sir Roger Wilbraham noted in his *Journal* that the Presence Chamber was the locale of the first night supper during the 1598 Feast. Assuming that this chamber was the favorite room for feasting at Garter celebrations at Whitehall, we may conjecture that if it were used for the Grand Dinner in 1597, there may not have been sufficient time to put it in order for a theatrical performance later that evening. This leaves us with a choice of the Great Chamber or the great Hall, either of which could have served for the occasion. Perhaps even the large and ornate Banqueting House may have been pressed into service. Although there are no records of a play presented in the Banqueting House during Elizabeth's reign, this hall had been built both for feasting and for various types of entertaining, including the presentation of masques. With the proposition presented in an earlier chapter that the accustomed supper on St. George's Night preceded the showing of the *Merry Wives*, the Banqueting House does appear as a definite possibility for the locale of the performance. (For further descriptions of the Banqueting House, see R. Holinshed, *Chronicles. . . ,* London, 1807-1808, IV, 434-435. See also Per Palme, *Triumph of Peace* [Stockholm, 1956], pp. 114-115, 129). The rolls of the Office of Works give no help in determining the appropriate chamber inasmuch as they contain no entries for any work done for the Feast of St. George in 1597. I surmise that whatever chamber was used, it contained a stage or raised platform from some previous occasion. Other than these tentative comments, the question of the exact locale for the presentation of the play can be pursued no further in the present study.

11 The preparations for this occasion, as described in the Revels accounts, are herewith reproduced from Albert Feuillerat, ed. *Documents*

1581 "at the Comaundement of the Lord Chamberleyne" for entertainment during the visit of the French ambassadors who had arrived in England to take up the matter of Elizabeth's marriage with the Duke of Anjou.[12] Again, in 1589, the Revels Office provided costumes "for six Maskers & six torchebearers. . . Sent into Scotland to the king of Scottes marriage by her Maiesties comaundement. . . ."[13]

The opportunity for finding a record of such expenditures for an "extraordinary" performance in 1597 has long

---

Relating to the Office of the Revels. . . (London, 1908), pp. 147, 153. The first entry refers to the "Ayryng, Repayrying [*sic*] Layeng abrode, Turning, sowinge, amending, Tacking, spunging, Wyping, Brushing, sweeping Caryeng, ffowlding, suting, putting in order and safe bestowing of the Garmentes, vestures, Armour properties, and other stuf, store & Implementes of the seide office for the safegarde Refreshing and Reddynesse therof & Agaynste the Coomyng of Duke Mommerancie Embassadour for ffraunce." And the second, for June 5, notes: "Woorkes doone & Attendaunces geven within the seide Office and on thaffares therof within the same tyme ffor & vpon the Devyzing, Newmaking [*sic*] Translating, Repayring, ffyttyng, ffurnishing, Garnishing, & setting foorthe of sundry kindes of Apparell properties and ffurnyture for One Maske showen at White hall before her Maiestie & Duke Mommerancie Embassador for ffraunce together with the Emptions & provisions bowghte and provyded for the same And all other Charges growen by meanes thereof within this Office (the Warderobe stuf as before is saide only excepted) perticulerly ensueth with the partyes names to whome any mony is due."

[12] "Attendaunce geven and worke done betwixt the xviij[th] daie of marche and the first of Aprill Anno predicto [1581] at the Comaundement of the Lord Chamberleyne for setting downe of paterns for maskes and making vp of some of the same for the Receaving of the ffrench Comissioners with the provision of certeyne stuffe properties and making of modells for a mownte and for the edifying of a greate parte of the said mounte The particularities wherof herafter ensueth. . ." (Feuillerat, p. 340).

[13] "Betwene the   of September 1589 *anno* xxxj[o] regni Regine Elizabethe and the   of the same September for newe making of dyvers garmentes & altering & translating of sondry other garmentes for the furnishing of a maske for six Maskers & six torchebearers and of such persons as were to vtter speches at the sheweng of the same maske Sent into Scotland to the king of Scottes mariage by her maiesties comaundement signified vnto the Master & other officers of this office by the Lord Treasorer, the Lord Chamberleyn & Mr vicechamberlene. The chardges aswell for workmanshipp & attendance as for wares deliuered & brought into this office for & about the same hereafter particulerly insueth" (Feuillerat, p. 392).

vanished since the extant accounts of the Master of the Revels cease with October 1589. But the above examples clearly illustrate that the Revels Office was fully equipped to meet the demands of "extraordinary" masques and plays even when only a single performance was given, and that the cost for furnishing the various items was met by the Revels Office unless some were loaned by the Wardrobe or the Office of Works or supplied by the company.[14] Nor were the expenditures for these functions small—that of the 1581 masque alone came to £730-18-6. Also worthy of note is the fact that the Master of the Revels frequently acted under direct orders of the Lord Chamberlain in preparing these "extraordinary" productions. There is no reason to doubt that similar procedures could have served to bring the *Merry Wives* before the expectant audience in 1597.

Thus far I have shown how Lord Hunsdon could have called upon the Usher of the Chamber and his assistants to prepare the hall for the performance as part of their duties at the Feast of St. George and how he could have summoned the Revels staff to provide costumes, properties, and scenic pieces for the production. Hunsdon likewise exercised control over the personnel at Court needed to fill out the performing company.

In addition to the actors, who, of course, would have come from Hunsdon's own troupe, the *Merry Wives* requires a group of singing boys and probably musicians to accompany them. Although the scanty stage directions in the Folio text militate against a definitive statement

14 It is the belief of Chambers that the drop in annual expenditures of the Revels Office after 1573 is caused by the Wardrobe and Office of Works assuming the supply of many needed items. See *Notes on the History of the Revels Office* . . . (London, 1906), p. 63; also *The Elizabethan Stage*, I, 90. Confirmation of this point of view is furnished by a passage in the Revels Accounts for 1572-73: ". . . saving the Warderobe stuf which is not here mencyoned bycawse it was not bowghte by any officer of the seide office but delyvered to thoffice by Iohn fforteskue esquier Master of the Queenes Maiesties greate Warderobe" (Feuillerat, p. 151).

about the need for musicians and the Quarto version, in addition to the stage directions for two songs, refers only to "a noise of hornes" (V.v. following line 35), the use of musicians seems logical. But whose musicians—the company's or those attached to the Court? Inasmuch as the need for music in the *Merry Wives* is minimal, it would seem a much simpler—and cheaper—matter if the court musicians were called upon. As the Treasury Accounts indicate, the court musicians were paid lump sums for their services on either a quarterly or an annual basis. Their attendance, therefore, at the Feast of St. George would have been part of their regular duties. Of the various bodies of musicians in the royal household one particularly stands out as suited for just such a task as the *Merry Wives* calls for: the group termed "Players of Enterludes" in the 1597 ". . . Collection of all the offices of England. . . ."[15]

If the musicians came from the royal household, may not the singing boys also be found there? Indeed, it is necessary to look no further than the Chapel Royal where about ten youngsters—highly trained in the performing

[15] "Musitions & Players" section in "A general Collection of all the offices of Englond with their fees in the Quenes guift viz. . . All the offices and ffees of her highnes Royall howshold with other Rewards and allowances . . . Coppied . . . in Aprill 1597," in B.M. Add. MS. 12,512, fol. 49. The entry for "Players of Enterludes" in 1597 is not an isolated one, for it appears in all the extant lists I examined: 1593, 1598, and 1600. Chambers cautions that manuscripts with this heading should not be regarded as official lists but as guides for courtiers seeking patronage (*The Elizabethan Stage*, I, 29).

This entire subject of the relationship between the court musicians and the dramatic companies is one about which comparatively little is known. Dr. John P. Cutts, who has investigated the subject from the standpoint of the Royal Musicians and the King's Men in *La Musique de la Troupe de Shakespeare* (Paris, 1961), concurs with my belief that probably the Queen's Musicians would have helped with productions of plays at Court during Elizabeth's reign. (Information in a letter dated January 8, 1958, from Dr. Cutts to me.) Of the various classifications of musicians in the royal household, it appears that those termed "Players of Enterludes" played only for court entertainments. Until further evidence is forthcoming, this conclusion can be considered only tentative.

arts—were maintained as choristers.[16] These children, more-over, took part in the Feast of St. George ceremonies.[17] Although they never received fees for their routine duties, the boys did get annual largess amounting to £9-13-4 for their attendance at high feasts.[18] If it happened that some of the Children of the Chapel Royal would not have been satisfactory to portray the fairies in the *Merry Wives*, they may have been replaced with any of the ten children from the Windsor Chapel choir inasmuch as the Chapel Royal and Windsor choirs have been known to form a united body for the St. George's Feast services.[19] Though it seems more probable that the Children of the Chapel Royal alone furnished the fairies, it is certain that a supply of about twenty talented youngsters was readily available at White-hall during the Feast of St. George. One doubts that much urging would have been necessary to get their enthusiasm worked up for this special production of the *Merry Wives*.

Though it has been possible to account for the prepara-tions and payment to those of the royal household who would have been concerned with readying this postulated production, no concrete evidence for payment to the actors is forthcoming. However, various theories suggest them-selves. Sometimes a play was rewarded from the Privy Purse rather than through the Treasurer of the Chamber,

16 C. C. Stopes, *William Hunnis and the Revels of the Chapel Royal* (London, 1910), pp. 20-21.

17 Ashmole, p. 574. For further information on the attendance of the Children and Gentlemen of the Chapel at the Feast of St. George, see the record of yearly benevolences for 1581-1667 in *The Old Cheque-Book . . . of the Chapel Royal. . .* , pp. 117-120.

18 Charles W. Wallace, *The Children of the Chapel at Blackfriars, 1597-1603* (Lincoln, Neb., 1908), pp. 3-4, 187; B.M. Add. MS. 12,512, fol. 52. This 1597 figure in "A general Collection of all the offices of Englond. . ." is duplicated in other such lists. See P.R.O. LC5/182, fol. 24v; B.M. Add. MS. 12,508, fol. 10v; Dulwich College MS. XI, fol. 19; Francis Peck, ed. *Desiderata Curiosa. . .* (London, 1732), I, 9, 60-61.

19 Ashmole, p. 576. In connection with these two choirs functioning as one, there is a Revels Accounts entry for a jointly presented *Mutius Scae-vola* at Court on January 6, 1577.

but since the only extant Privy Purse account is for 1559-69, this hypothesis cannot be pursued further.[20] Tracing through the Revels Accounts is also blocked by a lack of records. Then there is the observation of Chambers that the actors received no monetary reward for an "extraordinary" production.[21] Perhaps the Chamberlain's Men were willing to play gratis, dually honoring their patron's election to the Garter and their reassumption of the title Lord Chamberlain's Men; or perhaps they may have considered it sufficient reward to have witnessed part of the Garter ceremonies (as many Englishmen do today). One last possibility—and that the strongest of all—is that Lord Hunsdon rewarded the actors from his own purse, paying them the accustomed ten pounds they received for a regular Court performance.

Whether the actors would have had adequate rehearsal time need not be questioned if I have postulated the calendar of events properly. Shakespeare, as previously indicated, had at least three to four weeks available to write the play. Now, all that the Dennis accounts of the tradition state is that the *Merry Wives* was written in fourteen days (possibly even ten). In fact, in the most well known of his three passages on the time element, Dennis states that the Queen "commanded it to be finished in fourteen days; and was afterwards . . . very well pleas'd at the Representation." This word *afterwards* is indeed an important one, for *afterwards* can imply a time lapse between the date the play was completed and the date of performance. This allows for copying parts and rehearsing.

Here is a period of twenty-one to twenty-eight days available for executing the entire project. Let us estimate twenty-five. The writing of the play takes up about fourteen days (no impossible feat especially if, as we shall

---

20 Chambers, *The Elizabethan Stage*, I, 62-63, 214, n.1.
21 *ibid.*, I, 214-215, n.2.

later see, one is adapting an old play). This leaves either five or six days for copying parts and five or six for actual rehearsals. Aside from those actors who are blessed with being "quick studies," five or six days is ample for a trained actor to work up a part. His performance, while not masterful, would be technically proficient. Anyone who has been with an American summer stock company or an English weekly repertory troupe knows that this length rehearsal period is practically standard. In this connection, one may wonder at the prodigious memories of Henslowe's actors who, as the *Diary* shows, often had to master a new play in a week to twelve days. This in addition to changing repertory daily. Therefore we can see that the schedule of events does allow for Shakespeare's completing the *Merry Wives* in sufficient time for the actors to prepare their roles.

That all required personnel and equipment were readily available at Court for the initial performance of the *Merry Wives* seems a fair assertion. Still, the play as a play would have to be licensed for performance. There is no point in laboring the question of whether Edmund Tilney performed his customary function as Master of the Revels of passing on the script by having the actors present the *Merry Wives* before him or by examining the manuscript. Ample entries exist in the Revels records to indicate that the Lord Chamberlain could—and did—interfere directly in the administration of the Revels Office. And scholars generally agree now that this interference extended to granting licenses to plays.

The foregoing discussion reveals that the artificers and officials of the Revels Office, the Children of the Chapel Royal, the Queen's Musicians, the Usher of the Chamber and his assistants—all members of the royal household who would have been called upon to make the first performance of the *Merry Wives* possible—were directly responsible to the same official: Lord Hunsdon, the new Lord Chamberlain. With the fees for their services covered either by annual salary or by monetary rewards charged against the Feast, the Lord Chamberlain would not have had any budgetary problems in marshalling the aforesaid departments to prepare and present the production at Whitehall. As for the actors and playwright, Lord Hunsdon had only to turn to his troupe (one of the finest in London), feeing them from his own purse.

CHAPTER V ✦ THE TEXT

MONG the various problems surrounding the genesis of the *Merry Wives* is that of determining the nature of the text used at the postulated 1597 Garter performance. The matter is complicated by the fact that the earliest printed version of the play—the First Quarto of 1602—differs in several major respects from the Folio text.[1]

The Quarto text is considerably shorter than that of the Folio, containing 1,624 lines, whereas the Globe edition of the Folio text numbers 3,018 lines.[2] The Quarto completely omits five scenes. IV.i.; V.i., ii., iii., and iv. It also transposes two others, III.v. appearing before III.iv. Speaking parts for the boys have been cut. (In the scene at Q1, 134-237, the boy, though on stage, speaks no lines.)

Aside from greater length, the Folio text differs markedly from its Quarto counterpart in rendering many passages. Then, too, it has its own physical characteristics. Few stage directions appear in the F version. The only ones specified are the massed entries at the head of each scene; one direction for an entrance listed at the actual place of entry within the scene (V.v.41); and an *Exit* or *Exeunt* notation at the conclusion of each scene (except in III.i. and IV.iv.). Furthermore, the *Merry Wives* has been divided into acts and scenes. Inasmuch as many other plays in the Folio are similarly divided, this feature has little

---

[1] For an excellent description of the physical characteristics of the Folio text, see an unpublished dissertation (University of Virginia, 1956) by Elizabeth Brock, "Shakespeare's *The Merry Wives of Windsor*: A History of the Text from 1623 through 1821," I, 1-98.

[2] These figures are from Greg. Alfred Hart counts 1,419 and 2,634 respectively. David White gets 2,624 for the Folio, also based on a Globe count, and concurs with the Greg quarto count. William Bracy, using the parallel texts of the Bankside edition, registers 1,620 and 2,701.

relevancy except for the fact that all Folio plays supposedly based on Ralph Crane transcriptions share this trait.[3]

The very existence of these variant texts raises important questions in establishing the nature of the script used for the 1597 production. Which of the two versions represents more fully the original acting version? If neither, do both stem from a common original? Why should two versions have reached print? In what manner does each complement the other?

Textual critics have been wrestling with these questions ever since Alexander Pope made the first attempts at a critical evaluation of both versions in his 1725 edition of Shakespeare's *Works*. The search for answers has proceeded in two directions: 1) a line by line study of the Quarto and Folio—since editors early recognized that collation of both texts is necessary for creating an authoritative edition—to determine which contains the superior readings, and when confronted by a crux, to substitute the editor's own conjectural emendations; 2) an investigation of the two versions as separate entities in an attempt to determine whether each goes back to a common original; whether the Quarto is the original or a later reworking of the text which furnished the copy for the Folio; or whether the Folio text with modifications can be considered the more authentic version. (The modifications stem from script alterations dictated by playing conditions and from changes in topical lines between the time of the first performance and the 1623 printing.) These two textual approaches, of course, are not mutually exclusive even though various editors and critics of the *Merry Wives* have tended to emphasize one or the other.

Of paramount concern in the present treatise is the establishment of the Quarto-Folio-original script relationship rather than a consideration of linear variances stem-

---

3 For discussion of Ralph Crane, see below, pp. 102-103.

ming from grammatical, orthographical, compositorial, and similar differences. Such a study of text relationship depends, to a large extent, upon evaluation of theories advanced by both present-day and early critics. But since bibliographical scholarship has proceeded at such a pace during the past fifty years—forcing textual specialists to modify, revise, and discard yesterday's theories—it is pointless in the present study to resurrect the outworn hypotheses born in the salad days of textual criticism simply for their own sake. Accordingly, only those concepts from the past necessary for substantiating the theory of text presented in this chapter have been temporarily resuscitated. Otherwise, let them rest in peace—to be disturbed only by the textual historian.[4]

The point of attack for getting at this Quarto-Folio-original script relationship lies in a bold frontal advance on the Quarto. The entry of this Quarto in the Stationers' Register on January 18, 1602, by John Busby, together with its simultaneous transfer of copyright to Arthur Johnson provides the first record of the existence of the *Merry Wives*.[5] The simultaneous transfer is now recog-

---

[4] For the most complete study of the transmission of text between 1623 and 1821, see Brock's dissertation. An excellent summary of eighteenth century textual analysis also appears in William Bracy, *The Merry Wives of Windsor: The History and Transmission of Shakespeare's Text* (Columbia, Mo., 1952), pp. 9-17, hereafter cited as *History and Transmission*. Bracy further devotes Chapters II-VI to a discussion of modern critical opinion on the play. For a detailed discussion of the theories of the modern textual critics, see also an unpublished dissertation (University of Iowa, 1942) by David Manning White, "The Textual History of 'The Merry Wives of Windsor,'" hereafter cited as "Textual History."

[5]	Io. Busby.	Entred for his copie vnder the hand of mr Seton/A booke called. An excellent & pleasant conceited comedie of Sr Io ffaulstof and the merry wyves of windesor vjd
	Arthure Iohnson	Entred for his Copye by assignement from Iohn Busbye/A booke Called an excellent and pleasant conceyted Comedie of Sir Iohn ffaulstafe and

nized as standard procedure among stationers of the period and probably represents nothing more than a straight business proposition.[6] Johnson published the Quarto within the year, using Thomas Creede as his printer. Its title page is an elaborate one:

A/Most pleasaunt and/excellent conceited Co-/
medie, of Syr *Iohn Falstaffe*, and the/merrie
Wiues of *Windsor*./Entermixed with sundrie/var-
iable and pleasing humors, of Syr *Hugh*/the Welch
Knight, Iustice *Shallow*, and his/wise Cousin M.
Slender./With the swaggering vaine of Auncient/
Pistoll, and Corporall *Nym*./By *William Shakespeare*./
As it hath bene diuers times Acted by the right
Honorable/my Lord Chamberlaines seruants. Both
before her/Maiestie, and else-where./London/
Printed by T.C. for Arthur Iohnson, and are to
be sold at/his shop in Powles Church-yard, at
the signe of the/Flower de Leuse and the Crowne./
1602.

A Second Quarto, issued by Pavier, appeared in 1619, but it is primarily a reprint of Q1 and therefore of no editorial significance.[7] (In fact, all seventeenth-century editions of

---

the merye wyves of windsor   vjd
[Register C, fol. 78 recto]

Quoted by W. W. Greg, ed., *The Merry Wives of Windsor*, 1602 (Oxford, n.d.), Introduction. This is the 1956 reprint of the Shakespeare Association Quarto Facsimiles of 1939, and is hereafter cited as Shakespeare Association facsimile. The Arber transcript of the entry shows minor variations in format.

[6] W. W. Greg, *Some Aspects and Problems of London Publishing between 1550 and 1650* (Oxford, 1956), pp. 63-81; Leo Kirschbaum, *Shakespeare and the Stationers* (Columbus, Ohio, 1955), pp. 66-74.

[7] Q3 (1630), except for a few changes in spelling and punctuation, was printed from F1. The Folios of 1632, 1663, and 1685 also depend upon F1. For the editorial differences in F2, F3, and F4, see Brock, I, 115-171. For differences between Q1 and Q2, see W. W. Greg, ed., *Shakespeare's Merry*

the play subsequent to the publication of Q1 and F1 are basically reproductions of either of these two versions.)

First to study the nature of the Quarto was Alexander Pope who concluded that the version "compos'd at a fortnight's warning" must have been "the first imperfect sketch of this Comedy. . . . This which we here have [the Folio text which he printed], was alter'd and improved by the Author in almost every Speech."[8] The fact that Pope knew only the 1619 quarto is immaterial, since this quarto (Q2)—as just indicated—is more or less a duplicate of the 1602 copy. (Theobald drew attention to the existence of Q1 in his 1733 edition of Shakespeare [1, 223].) What does matter, however, is that Pope recognized that the Quarto was in some way a faulty text—a conclusion never disputed by later scholars. Subsequent eighteenth- and nineteenth-century editors such as Theobald, Johnson, Capell, Malone, Knight, and Halliwell-Phillipps were—with slight individual modification—content to accept Pope's theory that the Quarto antedated the Folio and represented a first sketch.[9]

The death blow to the "first sketch" theory was decisively dealt by P. A. Daniel in his introduction to the Griggs facsimile of the Quarto (1881). Noting that both the Q and F have textual deficiencies and that the Q version also renders certain readings superior to that of the F, Daniel postulated that both went back to a common original. He

---

Wives of Windsor: 1602 (Oxford, 1910), pp. 95-100, hereafter cited as 1910 Q1 facsimile.

8 *The Works of Shakespear* (London, 1725), I, 233.

9 This was the view of J. O. Halliwell-Phillipps when he prepared his introduction to the 1842 reprint of the Quarto; see *The First Sketch of Shakespeare's Merry Wives of Windsor* (London, 1842), pp. v-xxxii. Later he abandoned this theory in favor of a shorthand piracy hypothesis to account for the origin of the Quarto. See *Outlines of the Life of Shakespeare*, 2nd ed. (London, 1882), p. 103.

further recognized that the scenes not present in the Quarto could be cut from a longer version without interfering with the intelligibility of the plot.[10] Thus, he concluded: "The meagreness of the Q may be accounted for by the well known common practice of the stage of shortening plays for representation, and as omissions in it can be proved, this seems to me the more reasonable solution of the question."[11]

No longer does any scholar attempt to sustain a case for the Quarto as the original version—later revised and expanded—of the *Merry Wives*. Dover Wilson, the last prominent critic to have advocated primacy for the Quarto by developing with A. W. Pollard what came to be known as the continuous revision theory,[12] has, in the light of continuing investigations by other scholars, reevaluated his position and renounced the theory.[13]

Thus we return to the hypothesis that the shorter Quarto version of the *Merry Wives* can be traced to some sort of official abridgment of a fuller script for presentation on the London stage. (This fuller script, as will shortly be discussed, is the text behind the Folio version.) What

[10] *Shakespeare's Merry Wives of Windsor: The First Quarto, 1602* (London, 1888), pp. vi-vii, x, hereafter cited as Griggs Q1 facsimile.

[11] *ibid.*, p. vii.

[12] According to this theory, the "bad" quartos of *Romeo and Juliet, Henry V*, and the *Merry Wives* existed in some form before May 1593—that of the *Merry Wives* being *The Jealous Comedy*. These plays were then cut for the long provincial tour which Lord Strange's Men commenced in the spring of 1593 when the prolonged outbreak of plague kept the London playhouses shut. Between 1593 and the date of the subsequent reinstatement of the fuller version into the repertory, Shakespeare had revised these plays. See "The 'Stolne and Surreptitious' Shakespearian Texts," TLS, Jan. 9, 1919, p. 18; Jan. 16, 1919, p. 30; Mar. 13, 1919, p. 134; Aug. 7, 1919, p. 420; Aug. 13, 1919, p. 434.

[13] In the New Cambridge edition of *King Henry V* (Cambridge, England, 1947), he states that Pollard "was at the time with me trying out ideas, later abandoned, on the origin of the 'bad' Shakespearean quartos . . ." (p. 113). See also p. 112 of this edition.

proponents of this view, such as Alfred Hart,[14] initially fail to do is to establish a reason why the *Merry Wives* should have been shortened for London performance. Since it is a play of average length,[15] abridgment to reduce playing time does not suggest itself. Nor does a need for reducing personnel seem a valid reason, for the boys thus eliminated from the cast still would have had to be present during a London season for the other plays in the repertory.[16] Neither does the content, as perusal of the Folio text proves, contain anything that demands extensive hacking away. No plausible reason can be advanced to support a thesis that the play was cut for production during the London theatrical season.

In fact, even Alfred Hart's detailed studies into play abridgments on the London stage confute his own stand on the *Merry Wives*. Drawing upon contemporary allusions by playwrights and actors to two hours as the duration of an Elizabethan theatrical performance (excluding the jig), Hart has decided that all such references must be taken literally as an indication that plays on the London stage ran "not less than two hours and not quite three."[17] Reckoning from this time limit, he has decided that the average length of a play at the Globe or at Blackfriars would have been about 2,400 lines.[18] This figure correlates closely with the 2,447 average he finds for "204 extant printed plays with sound texts written by all authors other

[14] *Stolne and Surreptitious Copies: A Comparative Study of Shakespeare's Bad Quartos* (Melbourne and London, 1942), *passim*.

[15] Hart, who counts 2,634 lines in the F text, estimates 2,400 lines as the average length of an Elizabethan play. See *Shakespeare and the Homilies* . . . (Melbourne, 1934), pp. 96-118. J. Dover Wilson reckons 3,000 lines as average, a figure which corresponds almost exactly with the Globe line count of 3,018 for the play. See *The Two Gentlemen of Verona* (Cambridge, England, 1921; rev. 1955), p. 81 for the Wilson figure.

[16] White, "Textual History," pp. 125-134.

[17] *Shakespeare and the Homilies*, p. 104.

[18] *ibid.*, pp. 96-118.

than Jonson during the years 1590 to 1616."[19] Longer plays required cutting. Short plays as well went under the exciser's knife. Accordingly, *John a Kent and John a Cumber* went from 1,672 to 1,638 lines, and *Edmond Ironside* from 2,061 to 1,865 lines. That such pruning of the playwright's original script in the course of rehearsals is a justified reality even in the modern theater is an unchallengeable assertion. Thus, Hart's conclusion about script cutting is entirely valid; what went were "excursions into the arid realms of philosophy, sage reflections on life, conduct and character, over-much moralizing, unnecessary displays of learning and classical allusions, for the good reason that such topics were not concerned with the action and were of little interest to the majority of the auditors. For the same reasons similes, elaborated comparisons, overworked metaphors, excessive word-play, out-of-date topical references, iteration and amplification of ideas previously expressed were always in danger of suppression. A line or two in one place, a short passage in another, a long speech or even an episode might disappear."[20]

But let us apply his statistics directly to the *Merry Wives*. As previously noted, Hart gives 2,634 lines to the Folio text and 1,419 to that of the Quarto. (The fact that other scholars have arrived at different figures must now be ignored since we are dealing solely with Hart's own statistics.) A play the length of the F text would have taken approximately two and a quarter hours to perform—by Hart's estimate. The number of lines cut amounts to 1,215 or roughly 45 per cent of the total number in the play.[21] Now, nowhere in Hart's table can a cut of such magnitude be found for the London production of a play. The two longest on his list, *Richard II* and *The Captives*, register

---

[19] *ibid.*, p. 122.   [20] *Stolne and Surreptitious Copies*, p. 133.
[21] Using the figures of White, we find a cut of about one-third, a figure which corroborates the above calculation.

respectively cuts of 4 and 6 per cent. And the shortest, *John a Kent and John a Cumber* has had only 2 per cent of its lines cut. Application of Hart's own methodology, therefore, shows how completely untenable any theory is that advocates official script abridgment of the *Merry Wives* for performance on the London stage. The running time alone would be slightly over an hour, and the Elizabethan demanded his money's worth at the theater.

Abridged the Quarto is, but abridged for performance where—and in what manner? When H. C. Hart considered the problem in his introduction to the Arden *Merry Wives of Windsor*, he came to the conclusion that the Quarto was an authorized shortened version of the original text condensed "for special purposes, or for private representation, as for example, for compression into reduced time after court revels or banquets."[22] Basic agreement with this point of view has been expressed as recently as 1952 by William Bracy (*History and Transmission*, pp. 96-97).

This theory of official abridgment for court or other private presentation is constructed on an extremely wobbly framework.[23] Let us remember the nature of the Q text: flawed, corrupt (although not incomprehensible). And let us remember the reputation of the company supposedly presenting such a special performance: a leading theatrical troupe of the time, under the direct patronage of the Lord Chamberlain, and having as author of the *Merry Wives* the outstanding playwright of the London stage who was also a shareholder in the company. Let us remember too the type of audience at a private performance: not garlic-breathed groundlings, but sophisticated, literate courtiers. Is it possible that such a competent company would dare

[22] Arden edition (London, 1904), p. xiii.
[23] My criticism of the theory of official abridgment for court presentation is directed solely at this Q version of the *Merry Wives* and is not to be construed as an indictment of the generally accepted theory that plays were on occasion abridged for performance at court.

perform such an inferior script (especially assuming, as H. C. Hart and Bracy do, that the company authorized the adaptation from its longer and more perfect version) before such a select audience? If this refutation of official abridgment for Court performance not be thought sufficiently strong, consider an added piece of evidence. Every single reference to either the Court or the Order of the Garter—and these allusions are all favorable—has been stricken from the Q text. A strange procedure—deliberately going out of one's way to avoid complimenting those who were paying the bill for the performance.

Thus far the quest for the origin of the *Merry Wives* Quarto has led to a rejection of any theories positing the genesis of the Q before that of the fuller text which served as the F copy, as well as a rejection of theories of official abridgment of the longer version either for presentation on the London stage or for special private performance. There is still another view to be considered, one originally advanced by Pollard and Wilson, that the Q text might have been shortened for presentation during provincial tours.[24]

All objections raised above for producing a cut version in London either for the public or private stage vanish at the suggestion of provincial presentation. As Greg so convincingly points out in his *Two Elizabethan Stage Abridgements: The Battle of Alcazar & Orlando Furioso* . . . (Oxford, 1923), shortening plays for provincial tour was in no way an irregular procedure among Elizabethan acting companies. He cautions, however, that no two plays follow the same methodology in making abridgments.

This theory of provincial abridgment has now gained general acceptance. Indeed, the special conditions of touring would have had a direct effect on the structure of the

[24] "The 'Stolne and Surreptitious' Shakespearian Texts," TLS, January 9, 1919, p. 18.

scripts. Shortening to emphasize plot and to cut away embellishments which would not appeal to local audiences as well as excising roles to accommodate a smaller troupe would have been normal procedures in preparation for such tours.

The altering could have been done in two ways: (1) officially by the London companies especially for provincial tour, or (2) memorially reconstructed on tour in the absence of an authoritative prompt book as Greg has demonstrated with *Orlando Furioso*.[25] A third possibility exists: illegitimate memorial reconstruction by one of the provincial acting troupes which were in direct competition with the London touring companies. References in contemporary literature as well as citations in Murray's *English Dramatic Companies* supply incontrovertible evidence about these provincial companies. C. J. Sisson has shown us how one such troupe, Cholmeley's Players, was giving unauthorized productions of *King Lear* and *Pericles* in the provinces around 1609-10.[26] As these two plays illustrate, the leaders of Cholmeley's Players (the Simpsons), knew what works were current and popular in London and worth purloining.[27]

Into which of these three categories of provincial alteration does Q *Merry Wives* fall? The garbled nature of the Q text tends to militate against official abridgment in preparation for the tour, for it is inconceivable that the textual corruption—recognized ever since 1725 when Pope labeled Q a "first imperfect sketch"—could have taken

[25] *Two Elizabethan Stage Abridgements*, pp. 333-357.

[26] "Shakespeare's Quartos as Prompt-Copies. . . ," RES, XVIII (April 1942), 137-141. Crompton Rhodes pursued the matter into the 18th century by calling attention to provincial piracies of Sheridan's plays. See his articles in TLS: "The Early Editions of Sheridan: 'The Duenna,'" September 17, 1925, p. 599; "The Early Editions of Sheridan: 'The School for Scandal,'" September 24, 1925, p. 617; also "Some Aspects of Sheridan Bibliography," *Lib.*, 4th Ser., IX (1928), 233-261.

[27] In this case the purloining was done from the printed quartos of *Lear* and *Pericles*.

place if the Quarto had been adapted officially by a member of the company. While there is no way of knowing what control, if any, Shakespeare had over his scripts once they were turned over to the company for production, one wonders whether from his special position of dramatist-shareholder-actor, Shakespeare may not have had the power to supervise textual alterations in his plays. If so, would he have permitted some hack to maul his *Merry Wives* script even if the original were hastily composed and even if a touring version were being prepared? Allowing that Shakespeare had no say in the matter or did not care, would the Chamberlain's company have had such a complete idiot for an adapter who could not cut, copy, and later verify lines from the prompt book before him—no matter how fast he might have been working?[28]

In turning to category two, legitimate memorial reconstruction[29] on tour, we know that the Chamberlain's Men were in the provinces during the summer of 1597. If they had decided to add the *Merry Wives* to their repertory, since the play was a new work, and if for reasons already shown they did not have a previously cut book with them, or had lost it, they would collectively have had to reconstruct a prompt copy. If this were what had happened—

[28] Hardin Craig, whose study *A New Look at Shakespeare's Quartos* (Stanford, 1961) appeared too late for detailed commentary in this book, supports my point of view by stating that "the company would have used its best talent in the preparation of the stage version" and "a sound and skillful job of adaptation would have been done" (p. 68). Craig, however, is arguing for adaptation by Shakespeare's company for a provincial tour in the summer of 1597, a conjecture which I consider unfounded.

[29] The term "memorial reconstruction" has come to mean any form of reconstructing a text without recourse to the printed original or to a stenographic transcript of an actual performance. The term is used throughout this study in this manner without distinction as to reconstructing from notes by a spectator, reconstructing solely from the memory of one actor, or reconstructing from the collective memories of a company. Leo Kirschbaum gives an account of the various types of textual corruption which may be explained by the memorial reconstruction theory in "An Hypothesis concerning the Origin of the Bad Quartos," PMLA, LX (September 1945), 697-715.

and we are piling up our *ifs* at a tremendous rate—we should expect a better job of reconstruction than the Q version gives us, intelligible as that version is; for we would still be close to the original private production which itself may have been followed by public performances between April 23 and July 1597, when the London theaters shut down. I cannot deny that there is a possibility of a 1597 provincial production of the *Merry Wives* by the Lord Chamberlain's Men, but aside from playing one conjecture against another with nonexistent records, we can adduce other evidence to reject this possibility. Why should the company have waited until 1602 to publish the text when it had already released the other Falstaff plays—*1 Henry IV* in 1598 and *2 Henry IV* in 1600? Sixteen hundred, as Pollard has pointed out,[30] was a crucial year for the dramatic companies. They appear to have been in trouble, and were raising money by the sale of their plays. Yet no *Merry Wives* on the market. Furthermore, why did they eventually release the inferior Quarto if they could have given out the F text? And lastly, how do we account for the appearance of Nym's tag line "there's the humor of it" only in the Q text?[31] It is necessary to study the Q version all the way through from its origin in an acting version to its publication to understand the nature of its text, and we see that legitimate cutting and/or reconstruction by the Lord Chamberlain's Men for a 1597 provincial tour cannot satisfactorily carry through on the publication problems.

A more all-inclusive explanation starts with a consideration of the third category of alteration, memorial reconstruction by and for an unauthorized party. Greg, in his introduction to the 1910 type-facsimile edition of Q1, ad-

[30] *Shakespeare's Fight with the Pirates*, 2nd ed. rev. (Cambridge, England, 1920), pp. 41-42.
[31] For discussion of this point, see below, pp. 88-95.

vanced the theory that the actor playing the Host had memorially reconstructed the Quarto of the play. Accuracy of reporting of the Host's own part as well as the parts of the actors with whom he appeared on stage provided the basis for this identification. Greg did note some additional close correspondences between the Falstaff scenes in the Q and F, but concluded, "I see no justification for conjecturing two agents where one will suffice."[32] Thus he completely overturned all previous critical opinion about the source of corruption in the Quarto by attributing it to the work of an actor-pirate making a memorial reconstruction from a stage version of the play.

From the time that Greg made his initial study of the Quarto in 1910, scholars, continuing the work of textual comparison between Q and F, have tallied an impressive array of botched readings in the *Merry Wives* Quarto. These have been subsumed under three main headings by White:[33] (1) anticipation and recollection of dialogue, of which he finds thirty instances between the two versions (2) the nature of the verse structure, a study of which may be used to account for the differences in quality on the grounds that a pirate actor was capable of fusing fragments of the text from various parts of the play into regular blank verse as well as of using lines from other plays in which he acted to fill in where his memory failed him,[34] and (3) the presence of *non sequitur* passages, which may be attributed to sheer oversight by the actor-pirate. Careless printing in Creede's shop, adds White,[35] brings further corruption to

[32] 1910 Q1 facsimile, p. xli.
[33] "Textual History," pp. 58-78. Appendix iii of his study gives a complete listing of the anticipations and recollections.
[34] Harry R. Hoppe calls attention to three borrowings from *Romeo and Juliet* in "Borrowings from Romeo & Juliet in the 'Bad' Quarto of *The Merry Wives of Windsor*," RES, XX (April 1944), 156-158. Noting that the lines are uttered by three different characters, he believes that this gives support to the pirate-actor theory.
[35] "Textual History," pp. 195-215.

the Q text. Since few dramatists proofread their formes anyway,[36] the errors perpetrated upon an illegitimately obtained text would have even less chance of being discovered.

Is indeed unauthorized memorial reconstruction the key to the existence of the *Merry Wives* Quarto? Opposition to the theory still arises from such people as William Bracy, who, in a dogged effort to discredit Greg and the entire school of memorial reconstruction, advocates a return to the official abridgment thesis. But Bracy, solid as he is on the historical portions of his dissertation, goes wide of the mark in his textual analysis—especially in the way he refuses to acknowledge the seriousness of the corruption in the Quarto.[37] Devoting a minimal amount of space to his textual theory, Bracy traces the quarto cuts with infinite patience, but then dismisses what is left with an almost casual assertion that the similarities between the Quarto and Folio versions can be accounted for only by postulating official adaptation. Yet, it is not the similarities but the *differences*—in many instances minute—which must be accounted for in determining the origin and relationship of the two texts of the *Merry Wives*.

And we see from the results of people like Greg and White who have made such a study that as Kirschbaum has maintained, "the memorial reconstruction theory is the only single theory that can account for all the various kinds of corruption. . . ."[38] In this judgment he is borne out by James G. McManaway (*Shakespeare Survey*, VI, 169) and

[36] Fredson Bowers, "The Problem of Variant Forme in a Facsimile Edition," *Lib.*, 5th Ser., VII (December 1952), 269.

[37] *History and Transmission*, pp. 79-97. C. A. Greer attempts a weak support of Bracy in "An Actor-Reporter in the 'Merry Wives of Windsor,'" N&Q, CCI (May 1956), 192-194. Hardin Craig, an arch-foe of the memorial reconstruction theorists, also backs up Bracy in *A New Look at Shakespeare's Quartos*, pp. 67-75. Craig, unfortunately, also makes little attempt to account for the differences.

[38] "An Hypothesis concerning the Origins of the Bad Quartos," p. 703.

Clifford Leech (MLR, XLVIII [July 1953], 333-335). And Greg himself, forty-five years after he first advanced a theory of memorial reconstruction by a traitor-actor to account for the origin of the *Merry Wives* Q, reaffirmed his initial stand. Stating that the Q is based on the F version, deliberately shortened in the course of preparing the pirated text, Greg concludes that "this was intended for acting, presumably in the country, and only later and incidentally found its way to press."[39]

The company performing this illegitimate version in the country was probably one of the provincial acting troupes not averse to getting hold of a recent London success. Now from the Q title-page legend "As it hath bene diuers times Acted by the right Honorable my Lord Chamberlaines seruants [,] Both before her Maiestie, and else-where," and from later stage records, including topical allusions from other plays and the 1619 quarto reprint of the *Merry Wives*, it appears as if the play became very popular soon after it entered the repertory of the Lord Chamberlain's Men. Therefore, the conditions under which Cholmeley's Players sought their scripts seem applicable to those under which some earlier provincial management would have tried to obtain an unauthorized *Merry Wives*; it was a recent and popular play written by the leading London dramatist.

In summation, then, the most probable explanation for the origin of the *Merry Wives* Quarto rests upon a combination of the memorial reconstruction theory with the data about unauthorized performances of London productions by provincial troupes.

When the memorial reconstruction was made is suggested by a piece of internal evidence in the Quarto. There is a verbal echo from *Henry V* which by reiteration establishes itself as a deliberate insertion in the illegitimate

[39] *The Shakespeare First Folio* (Oxford, 1955), 334.

Quarto—filched in the vein of the character who utters it. I refer to Nym's famous tag line "And there's the humor of it."

Nym utters this line five times in the Quarto, yet F *Merry Wives* has no counterpart for it as the following comparison of passages shows.[40]

Q Text	F Text
1. *Nym.* Syr my honor is not for many words, But if you run bace humors of me, I will say mary trap. And there's the humor of it. (lines 57-59)	*Nym.* Be auis'd sir, and passe good humours: I will say marry trap with you, if you runne the nut-hooks humor on me, that is the very note of it. (I.i.170-73)
2. *Nym.* His minde is not heroick. And theres the humor of it. (ll. 162-63)	*Ni.* He was gotten in drink: is not the humor cōceited? (I.iii.25-26)
3. *Nym.* Here take your humor Letter againe, For my part, I will keepe the hauior Of reputation. And theres the humor of it. (ll. 211-13)	*Ni.* I will run no base humor: here take the humor-Letter; I will keepe the hauior of reputation. (I.iii.85-86)
4. *Nym.* With both the humors I will disclose this loue to *Page.* Ile poses him with Iallowes, And theres the humor of it. (ll. 228-30)	[The Quarto is corrupt in reconstruction, combining Nym's lines at I.iii.103-104 with those at I.iii.109-12 which end with "that is my true humour."]
5. *Nym.* Farewell, I loue not the humor of bread and cheese: And theres the humor of it. (ll. 374-75)	*Nim.* . . . adieu, I loue not the humour of bread and cheese: adieu. (II.i.140-41)

Where did the pirate get the inspiration to employ this line in his memorial reconstruction, a line in each instance turned into a regular tag line? The utterances of Nym in

---

40 Of the passages cited below, Example 2 was edited into the Folio text by Theobald, but most modern editors have rejected the reading. Capell edited Example 5 into the Folio text; his eighteenth century successors as well as all modern editors have accepted the emendation.

Henry V, which appeared in 1599, suggest themselves. But let us look closely at the line as it appears in the F text of that play.

Shakespeare again places it in Nym's mouth five times, yet the phraseology is not the same as that of the tag line in the *Merry Wives* Quarto. It appears thus in *Henry V*:

II.i.63   "and that's the humor of it"
II.i.74   "that is the humor of it"
II.i.101   "that's the humor of it"
II.i.121   "Well, then that [*sic*] the humor of it"⁴
II.iii.63   "that is the humor of it"

The difference between "there's the humor of it" and either "that's the humor of it" or "that is the humor of it" is not particularly great, and may seem to be explained away as an alteration made by the actor in preparing his *Merry Wives* script. In fact, the "there's the humor of it" reading is one that falls easier on the tongue and is the type of change that a player would readily make.

However, before accepting this explanation, we must examine the same line in Q1 of *Henry V*. And here we find the startling fact that the tag line appears eight times in the Q and in the form "and there's the humor of it." These are the parallel passages.⁴² (The asterisks before numerals indicate passages with the tag line appearing only in Q.)

Q Text	F Text
1. Nim. I dare not fight, but I will winke and hold out mine Iron: It is a simple one, but what tho; it will serue to toste cheese, And it will endure cold as an other mans sword will,	*Nym.* . . . .I dare not fight, but I will winke and holde out mine yron: it is a simple one, but what though? It will toste Cheese, and it will endure cold, as another mans sword will: and there's an end. (II.i.7-11)

⁴¹ The error in this line probably stems from the compositor.
⁴² *Henry the Fifth: 1600.* Shakespeare Quarto Facsimiles No. 9, ed. W. W. Greg (Oxford, 1957).

And theres the humor of it.
(Sig. B1, 29-32)

*2. *Nim.* I must do as I may, tho
patience be a tyred mare,
Yet sheel plod, and some say
kniues haue edges,
And men may sleepe and haue
their throtes about them
At that time, and there is
the humour of it.
(Sig. B1v, 1-4)

Nym. I cannot tell, Things must be as they may: men may sleepe, and they may haue their throats about them at that time, and some say, kniues haue edges: It must be as it may, though patience be a tyred name, yet shee will plodde, there must be Conclusions, well, I cannot tell. (II.i.22-28)

3. *Nim.* . . . If you will walke off a little,
Ile prick your guts a litle in good termes,
And there's the humour of it. (Sig. B2, 1-3)

Nym. . . . If you would walke off, I would pricke your guts a little in good tearmes, as I may, and that's the humor of it. (II.i.61-63)

4. *Nim.* Ile cut your throat at one time or an other in faire termes,
And there's the humor of it.
(Sig. B2, 11-12)

Nym. I will cut thy throate one time or other in faire termes, that is the humor of it. (II.i.73-74)

5. *Nim.* That now I will haue, and theres the humor of it.
(Sig. B2, 33)

Nym. That now I wil haue: that's the humor of it.
(II.i.101)

6. *Nim.* Why theres the humour of it. (Sig. B2v, 11)

Nym. Well, then that [*sic*] the humor of't. (II.i.121)

7. *Nim.* I cannot kis: and theres
the humor of it.
(Sig. C1, 13)

Nim. I cannot kisse, that is the humor of it: but adieu.
(II.iii.63)

*8. *Nim.* Tis honor, and theres the
humor of it.
(Sig. C2v, 12)

[Q is corrupt in reconstruction; F III.ii.5-6 reads for Nym, "the humor of it is too hot, that is the very plaine-Song of it."]

We note in this comparison the same phenomenon that we did in the *Merry Wives* passages. The phraseology in the Folio version is varied from that of the Quarto, even though a form of the tag line appears in the F text in *Henry V*, but the Quarto tag line has been regularized and always placed at the end of the speech. That Shakespeare never thought of regularizing and positioning it thus in the authoritative F version is clearly indicated by noting Pistol's speech at III.ii.7-8, which picks up Nym's "the humor of it is too hot, that is the very plaine-Song of it" (example 8, F text, above) with "The plaine-Song is most iust: for humors doe abound. . . ." Here again we are faced with divergence in text between the Quarto and Folio.

Scholarly consensus recognizes that Q1 of *Henry V* is a "bad" quarto, an inferior and shortened version of the Folio text. It also appears to be a memorially reconstructed text. Whether it represents a reported text illegitimately reconstructed for provincial touring from the already shortened version or one authoritatively cut for such a tour cannot be determined. Greg leans toward the latter view.[43] There are other complicated theories on the subject, but for our purposes we need not pursue the textual relationships between the Q and F versions of *Henry V* any further.

What does concern us, however, is the publication of the *Henry V* Quarto in 1600 by Thomas Millington and John Busby. This had to take place between August and December. That it had not occurred earlier in the year is proved by the August 4 staying order for *Henry V* in the Stationers' Register and also by the entry for transfer of copyright to Thomas Pavier on August 14 in clear violation of the staying order. One of the publishers, John Busby, is the same individual who entered the illegitimate *Merry Wives* Quarto in the Stationers' Register on January 18, 1602.

It is possible, therefore, that the same actor—probably a "hired man"—was involved in supplying Busby with the texts of the two plays. With the *Henry V* Quarto he may merely have functioned as an agent in obtaining the shortened memorially reconstructed version. Possibly he had even participated in the authorized cutting session—

[43] See *The Shakespeare First Folio*, p. 282.

if we accept the Greg hypothesis—so that he had gained experience in the craft of adapting and reconstructing. If so, he apparently remained undetected in his dealings with Busby and Millington.

Some time the following year this same actor—who had been assigned the role of the Host in the 1597 production of the *Merry Wives*—applied his lessons well and made the illegitimate reconstruction of that play. White, who also favors a 1601 date for the piracy,[44] conjectures that this actor was released from the Lord Chamberlain's Men in early 1601 because the company was going through difficulties at the time—competition from the children's theaters, the War of the Theaters, harassment from the Puritans. He then joined a touring company for whom he memorially reconstructed the *Merry Wives*.[45] If White is correct, in the process of preparing his script for this company, the pirate followed the practice of whoever was responsible

---

[44] "Textual History," pp. 89-92.

[45] A tempting theory about this actor has been advanced by Henry David Gray in an article "The Rôles of William Kemp," MLR, XXV (July 1930), 261-273. Gray cites (on pp. 271-272) passages from the conclusion of Will Kemp's *Nine Days Wonder* (1600) in which Kemp seeks to find the anonymous ballad-maker who has been attacking him. In his search, Kemp narrates that he encounters "a proper upright youth, only for a little stooping in the shoulders, all heart to the heel, a penny poet, whose first making was the miserable stolen story of Macdoel, or Macdobeth, or Macsomewhat." This youth directs Kemp to another ballad-maker. The second ballad-maker, upon learning of the youth's accusation, merely laughs it off, noting that this "penny poet" is "a hoddy-doddy, a hobble-de-hoy, a chicken, a squib, a squall, one that hath not wit enough to make a ballad, that, by Pol and Aedipol, would Pol his father, Derick his dad, do anything, how ill soever, to please his apeish humor. I hardly believed this youth that I took to be gracious had been so graceless; but I heard afterwards his mother-in-law was eye and ear witness of his father's abuse by this blessed child on a public stage, in a merry Host of an Inn's part." Here then, according to Gray, is an actor-poet indentified by a sometime member of the Chamberlain's company as having played "a merry Host of an Inn." As a "penny poet" his talent would have been equal to turning out the wretched verse of the *Merry Wives* Quarto as well as supplying prose whenever his memory failed him. If Kemp were not writing vindictively, we would further have an actor not adverse to stooping to piracy within a year of the appearance of *Nine Days Wonder*.

for regularizing Nym's tag line in the *Henry V* Quarto from the F "that is the humor of it" to "there's the humor of it," and the pirate did the same with Nym's lines in the *Merry Wives*.

Allowing that the above conjectures may be too freely made, we are still left with these facts in dating the memorial reconstruction of the play. The Folio text version (1597) does not contain the "there's the humor of it" lines. The Folio version of *Henry V* (1599), which, according to Greg is based on Shakespeare's foul papers, reads "that's the humor of it." The *Henry V* 1600 Quarto is the first to preserve the reading "there's the humor of it." Thus it must be from this Quarto that the pirate of the *Merry Wives* obtained the reading that he placed in the 1602 Quarto of this play. If he were in no way connected with the reconstruction of *Henry V*, he simply could have consulted its printed Quarto to refresh his memory on the character of Corporal Nym.

That the *Merry Wives* reconstruction may have been made either before the stage appearance of *Henry V* or before the printing of its Quarto, that is between the summer of 1597 and the summer of 1599 or between 1599 and 1600, is improbable. This would mean in the first instance that Shakespeare would have somehow gotten hold of the line from the pirate and then used it in a debased manner in writing *Henry V*. Even if the line were a common catchphrase of the day, we could not explain away its use for the same character in practically the same way as sheer coincidence. Furthermore, why would the pirate have delayed so long in getting the play into print? Similarly, in the second instance—between 1599 and the appearance of Q *Henry V*—if the *Merry Wives* were such "hot property," why should the pirate have waited two years for its publication? After all, the other Falstaff plays had recently been published: Q1 and Q2 of *1 Henry IV* in 1598 and 1599 re-

spectively, and Q1 of 2 *Henry IV* in 1600. The fivefold appearance of Nym's tag "there's the humor of it" in Q1 of
the *Merry Wives* therefore presents strong evidence for
dating the reconstruction no earlier than the fall of 1600
but probably sometime in 1601.

There is a good possibility, we may speculate, that the
play had been revived about this time since in addition to
the quarto publications cited above, the production of
Henry V in 1599 and the unauthorized publication of
that play in 1600 may have created a great public demand for the Falstaff plays. Reference to "the swaggering
vaine of Auncient *Pistoll*, and Corporall *Nym*" on the title
page of the *Merry Wives* presents a further indication that
the public had developed a predilection for the Falstaff
crew upon the appearance of *Henry V*, for the roles of Nym
and Pistol in the *Merry Wives* are scarcely long or important enough to warrant such advertising in their own
right. Certainly it would seem necessary to have had a
performance of the comedy shortly before the piracy, for
it is doubtful whether an actor in the Elizabethan theater,
where plays had short runs with revival at irregular intervals, could otherwise have retained a working knowledge
of the complete script over a long period of time.

In making his reconstruction, the pirate would naturally
have cut away all material which might seem extraneous to
a provincial audience. Indeed, most critics have commented
that with all its garbling, the Q text is generally a coherent
one with the emphasis on a swiftly flowing plot. [46] The cutting of the opening scene in which Shallow's coat-of-arms
is described is but one sample of such textual pruning. Of
even more importance is the fact that all references to
either the Court or the Order of the Garter have been ex-

---

[46] This point of view led Vincent H. Ogburn to suggest that the pirate-
adapter was an expert in the structure of interludes. See "The Merry Wives
Quarto, a Farce Interlude," PMLA, LVII (September 1942), 654-660.

cised from the Quarto, although the Folio has eight such passages. Five of these have already been cited in Chapter One: I.iv.54, 61-62, 130; V.v.47-50 and 59-77. The first three state that Caius is on his way to the "grand affair" at Windsor Castle; the fourth contains the instructions to the fairies to clean the castle and ends with the implied salute to Queen Elizabeth; the fifth is the tribute to the Order of the Garter. In addition to these five, there is a passage at II.ii.62-81 in which Mistress Quickly mentions that "when the Court lay at *Windsor*" many of the noblemen were suitors to Mistress Ford. Two remaining passages—II.iii. 95-97 and IV.iv.88-89—comment on the influence of Dr. Caius in the community and show why his invitation to the "grand affair" may be considered a proper one.

While it is possible that in making the memorial reconstruction the actor playing the Host simply forgot some of these passages, especially those one or two lines long, the fact that all have been cut from the Quarto seems to indicate that the reconstructor worked with deliberate design in eliminating most (if not all) of them. His reason for doing so is clear. Not one of the Court–Garter passages is essential to the action of the *Merry Wives*. Intent on making a taut plot for provincial presentation, the actor–play doctor simply omitted the extraneous passages, most certainly knowing why they had been placed in the text originally. Their sole presence in the Folio version argues for the point advanced in this study that the *Merry Wives* was especially written for a Garter Feast.

It does seem odd that the Host-"pirate" should have cut these passages but not the horse-stealing fragment. That episode adds nothing to the intrinsic playability of the script. In fact, in its Quarto version it is so hopelessly corrupt as to be virtually meaningless. One possibility suggests itself for the retention of the episode. At the time of the reconstruction, as a result of the difficulties with the

Hanse, Germans were still unpopular in England.[47] Reference to a cosening German duke (as the Quarto text has it) would sit just as well with a provincial audience as with the Londoners. So for the sake of satire and topicality, the horse-stealing subplot is retained in the Quarto, but the gingerbread of Court and Garter passages gets cut away.

The pirate would also have eliminated as many non-essential roles as possible. Thus the excising of the roles of William Page and Robin would have been consistent with the dramatic practice of playing with reduced companies in the provinces. The pirate further removed some of the anachronisms in the original text. White, for example, analyzes the variant readings in the color of Anne's fairy scene costume on these grounds.[48]

Whether the actor playing the Host was the sole pirate is a question which has been reopened by the original champion of this view. Greg, in *The Editorial Problem in Shakespeare*, states: "Even apart from the two scenes dealing with the horse-stealing (in which the fact that the Host did not know his own part might be due to its having been altered) the superiority of the scenes in which he appears is not quite uniform (for instance the beginning of II.ii, when he is off, is better than the end of II.i, when he is on): moreover, there are errors in his speeches that perhaps suggest mishearing rather than the blunders of a compositor (IV.v. 93, 'I am cozened Hugh, and coy Bardolf' for 'hue and cry'!). Perhaps it would be safer to assume an

---

47 See below, Chapter VII.

48 "Textual History," pp. 82-87. In F V.v. the stage directions (New Cambridge edition) read, "Doctor Caius . . . steals away a fairy in green; Slender . . . a fairy in white; and Fenton comes, and steals away Mistress Anne Page." *How did he recognize her?* The Q stage directions in this scene read, ". . . the Doctor . . . steales away a boy in red. . . . And Slender . . . takes a boy in greene: And Fenton steales misteris Anne, being in white." There is still some internal inconsistency, but ultimately three distinct costumes appear in the Q version.

independent reporter relying generally on mine Host's assistance."[49]

In his latest pronouncement on the subject in *The
Shakespeare First Folio*, he elaborates his opinion still
further, and, going back to an original observation of his
1910 study, notes that both the Host's and Falstaff's parts
are highly accurate: "In neither case, however, is the quality of the reporting uniform, and it seems less likely that Q
was vamped up by the two actors themselves than that an
independent reporter was able to draw on the recollection
of each." In the mangled last act of the Q version the Host
is completely absent from the stage. Greg doubts whether
even the presence of Falstaff were enough for the independent reporter to re-create the scene; so he rewrote it.

The difficulty with bringing in the second actor, the
one who portrayed Falstaff, as an accomplice is that this
presupposes that the same actor played the part with the
Chamberlain's company. Now Falstaff, the leading role,
would not have been portrayed by a "hired" man. In fact
critics, notably T. W. Baldwin in *The Organization and
Personnel of the Shakespeare Company* (Princeton, 1927),
have attempted to identify the role with either Will Kemp
or Thomas Pope.[50] Neither of these seems a likely candidate to have consorted with some unknown "hired" actor
in an off-beat provincial company piracy of a play belonging to a company in which they had a vested interest. It is
far more likely that the actor-pirate either worked alone
or with Greg's unidentified independent reporter. In the
absence of proof positive, we can but conjecture.[51] In no

[49] 2nd ed. (Oxford, 1951), p. 71.

[50] See also T. W. Baldwin, "William Kemp not Falstaff," MLR, XXVI
(April 1931), 170-172; and two articles by Henry David Gray: "The Rôles
of William Kemp," MLR, XXV (July 1930), 261-273; "Shakespeare and Will
Kemp: A Rejoinder," MLR, XXVI (April 1931), 172-174.

[51] Though Kemp left the company in 1599, the allusion in *The Return
from Parnassus* to his trying out new talent with Burbage has been taken
as an indication that he briefly rejoined the Chamberlain's Men. A refer-

way, however, is the validity of the illegitimate memorial reconstruction theory challenged.

But how did this makeshift prompt book of the *Merry Wives* get into print? White[52] believes that the actor returned to London when the touring company disbanded in late November or early December 1601, with the onset of winter; he then sold the manuscript to Busby in much the same manner as Greg describes the sale of the *Orlando Furioso* prompt book. Pollard had indicated many years before that "the 'hired men' in the Elizabethan theatres were poorly paid, and still more poorly esteemed, and if one of them made up his mind to add to his earnings in this fashion, it might have been some time before detection overtook him."[53]

Corroboration for the 1601 date of the piracy is further found in an examination of Pollard's account of entries of plays in the Stationers' Register for the years 1591 to 1605.[54] With the exception of the period October 1593 to July 1594—a plague year—and the year 1600—one fraught with Puritan restrictions on the theater—few plays were entered in the Register. In the two above-mentioned years we find twenty-eight for the former and twenty-two for the latter. Pollard interprets this to mean that the plays were released legally because the companies were under duress and had to raise money. In 1601, eight plays were entered, and in 1602, three. Thus we see that once the

---

ence—cited by Halliwell-Phillipps and other critics—to a Latin note written by William Smith of Abingdon (B.M. Sloane MS. 392, fol. 401) dates this event prior to September 2, 1601. Assuming the accuracy of this interpretation, any case for tainting Kemp as a pirate at the same time he was a member of the Chamberlain's Men appears further damned.

As for the possibility of the actor-pirate working alone, both Greg and Kirschbaum report successful experiments in memorial reconstruction by one individual. See 1910 Q1 facsimile, pp. xxvii-xxix and Kirschbaum's "An Hypothesis concerning the Origin of the Bad Quartos," pp. 705, 708.

[52] "Textual History," p. 186.

[53] *Shakespeare's Fight with the Pirates*, p. 40.   [54] *ibid.*, pp. 41-42.

London companies had regained a stable position, it no longer became necessary for them to sell their plays. But a hungry actor, back from a provincial tour with an unauthorized manuscript of a popular play under his arm and faced with a cold winter in the capital, would be sorely tempted to cash in on the popularity of his play to tide him over a slack season. And there was the publisher Busby, hovering buzzard-like over the London dramatic scene—fresh from his plunder of *Henry V* the previous year, ready for another fat Falstaff play.

Assuming the above hypothesis of how the unauthorized and shortened acting version of the *Merry Wives* got into print to be a sound one, we are faced with the question of why the Chamberlain's Men took no steps to counter the publication of this debased *Merry Wives* Quarto by issuing an authoritative copy as they did with other bad quartos. Copyright procedures of the Stationers Company furnish the answer. Once a stationer established a copyright in a work—*no matter how the copy was obtained*—he had absolute control of that work; thus it was practically impossible for a legitimate owner, such as a dramatic company, to replace the corrupt text with an authoritative one once the former had been entered in the Stationers' Register.[55] Greg cites evidence on this point from George Wither, who in his *Scholars' Purgatory* (ca. 1624) writes of the typical stationer: "If he get any written copy into his power likely to be vendible, he will publish it; and it shall be contrived and named according to his own pleasure, which is the reason so many good books come forth imperfect and with foolish titles."[56] On the basis of these remarks and on other

[55] Greg, *Some Aspects and Problems of London Publishing*, pp. 63-81; Kirschbaum, *Shakespeare and the Stationers*, pp. 56-61, 65, 131-141, 207-209. Greg and Kirschbaum disagree on whether the copyright owner of a bad text owned all rights to the work or only to the particular bad quarto he published. Greg advocates the latter view.

[56] Quoted by Greg, *Some Aspects and Problems of London Publishing*, p. 75.

case histories, Greg states that "it may be significant that bad Quartos of *Hamlet, Romeo and Juliet,* and probably *Love's Labour's Lost,* which had been printed either without entrance or in violation of one, were all promptly replaced by authoritative versions, whereas similar quartos of *The Merry Wives,* which had been regularly entered, and of *Henry V,* to the copyright of which a plausible claim might be made, remained unsuperseded till the appearance of the collected folio of 1623."[57]

With the publication of the First Folio, then, the more authentic version of the *Merry Wives* finally got into print. This text, however, is an imperfect one. Recognition of this fact led Daniel and H. C. Hart to postulate a common original for the Folio and Quarto versions. What they failed to recognize in their comparatively early attempts to reconcile the two versions is that the Q text represents a memorially reconstructed acting version made by a player conversant with the original text and that this original went through minor changes between its composition in 1597 and its ultimate appearance in 1623. Such revision can be expected to take place during a twenty-odd-year playing span for a popular comedy, particularly with regard to topical references or "local jokes." Chief among the textual changes is the excision of oaths in accordance with the 1606 "Act to Restraine Abuses of Players."[58] Other minor changes in text—but of major importance in dating the composition of the *Merry Wives*—include the Brooke-Broome name change and alterations in the horse-stealing subplot. These variances are discussed in detail in the following three chapters.

Divergence between the Q and F texts can therefore more properly be explained by the fact that certain alter-

[57] *ibid.,* p. 74.
[58] See 1910 Q1 facsimile, Appendix ii, pp. liv-lvi, for a comparison chart of the oaths in the Q and F texts.

ations—such as *garmombles* to *Cozen-Iermans*—had been made in the original text after 1601, the posited date of the piracy; that certain other alterations—such as *Brooke* to *Broome*—had been made especially for the 1597 production and these the traitor-actor restored as he later made his memorial reconstruction; and that still others—such as the elimination of confusion in costume color in V.v.— had been placed directly in the provincial company script as the pirate made corrections based on his own playing knowledge of anachronisms present in the hastily composed 1597 text. Thus there is no basis for the supposition that both the Q and F texts stem from a common original. The F text is, with minor modifications, the authoritative version of the *Merry Wives*, and basically represents the script played at the 1597 Feast of the Garter.

What stood behind the copy furnished the printers of the Folio is difficult to determine. In the *Shakespeare First Folio* (pp. 336-337) Greg affirms that there is no concrete evidence to decide between the foul papers or a prompt book—especially since the absence of stage directions in the Folio text further complicates matters.[59]

In addition to the paucity of such directions, the F *Merry Wives* exhibits other peculiar characteristics which have led Greg to suggest that the printer's copy came from a literary transcription, probably by Ralph Crane. Greg believes that Crane, who became scrivener for the King's Men in 1619, made such transcriptions for the first four plays in the Folio (of which the *Merry Wives* is third) and for *The Winter's Tale* as part of an original plan by the Folio compilers to prepare the entire volume from literary

[59] At one time this lack of stage directions coupled with the collective grouping of the names of the characters at the head of each scene led Crompton Rhodes to postulate a theory that the copy for the Folio must have been reconstructed from the "plot," and by pasting together the individual players' parts. This "assembled text" theory is no longer tenable. (See 1955 reprint of the New Cambridge *Two Gentlemen of Verona*, p. 82.)

transcriptions.[60] The *Merry Wives* contains the very features which scholars have established as those belonging to scribal transcriptions: punctuation peculiarities, collective grouping of character names at scene beginnings, a scarcity of stage directions, and division into acts and scenes.[61] Thus a literary transcription as the immediate source of the Folio copy for the *Merry Wives* takes us still further from the original text, and demonstrates again the complex problems encountered in attempting to re-create the script used for the 1597 Garter production.

The Folio version, let me reiterate, does bring us close to that original text, sufficiently so to declare the F text authoritative. While minor alterations attributable to playing conditions do account for some slight differences between the F and mother texts, we must not overlook the fact that not all textual difficulties can be blamed on transmission; for the structure of the *Merry Wives* is flawed, and therefore several imperfections in the text are traceable to faulty craftsmanship by Shakespeare. Hasty composition can account for many of the repetitions in the script, particularly those found in the fourth and fifth acts; similarly, anachronisms in the time sequence in Act III and in the colors of the costumes in the fairy scene indicate a hurried job; the incomplete Justice Shallow–Falstaff quarrel and the fragmentary horse-stealing subplot point to writing without revising. Turning again to transmission of text, we note that transcription from either a prompt book or

---

60 *The Editorial Problem in Shakespeare*, p. 141. Ronald B. McKerrow shares this view in an article "A Suggestion Regarding Shakespeare's Manuscripts," RES, XI (1935), 459-465.

61 See *The Winter's Tale*, eds. J. Dover Wilson and Arthur Quiller-Couch (Cambridge, England, 1931), pp. 113-119; F. P. Wilson, "Ralph Crane, Scrivener to the King's Players," *Lib.*, 4th Ser., VII (1926), 194-215; R. C. Bald, ed. *A Game of Chess*, by Thomas Middleton (Cambridge, England, 1929); R. C. Bald, " 'Assembled' Texts," *Lib.*, 4th Ser., XII (1931), 243-248. For a detailed analysis of the manner in which Crane's scribal idiosyncrasies are reflected in the *Merry Wives*, see Brock, I, 2-13.

the foul papers suggests another area for contamination, for even the best of scribes was not infallible.

And, lastly, the actual printing of the Folio in the shop of Jaggard allows room for further mistakes to creep in. It is now known that several compositors worked on the book between 1621 and 1623 and that these varied in their accuracy and in the types of errors they made.[62] Furthermore, Jaggard's compositors set the Folio into type by the casting off method. Charlton Hinman, discoverer of this fact, sums up its importance: "For there are many indications that the Folio compositors were quite accustomed to expand or compress their type-pages in response to space requirements established by casting off; and *there are also indications that the adjustments which they effected sometimes involved altering the text of the copy.*"[63]

It now becomes possible to reconstruct the logical course of events which took place between the writing of the play and its ultimate appearance in the First Folio, events which vividly demonstrate why it is so difficult to determine the status of the script which I believe was used at the 1597 Garter Feast performance. As I have postulated in previous chapters, the play was commissioned on short notice for this special occasion. Both the pressure of writing with little time available and the request of the Queen to see Falstaff in a romantic entanglement placed a tremendous burden upon Shakespeare in terms of planning his plot structure. His shortcomings are evident, and they initially mar the script.

But what emerged from his pen for the 1597 production may be considered the sole authentic text of the play. Ex-

[62] See Brock (I, 37-42) who discusses and expands an analysis made by Irby B. Cauthen, Jr. of compositorial assignments for the *Merry Wives* in an unpublished dissertation (University of Virginia, 1951) "Shakespeare's *King Lear*: An Investigation of Compositor Habits in the First Folio and their Relation to the Text."

[63] "Cast-off Copy for the First Folio of Shakespeare," SQ, VI (Summer 1955), 263. Italics mine.

cept for minor revisions, it served as copy for the Folio. About 1601, however, a "hired" man of the Chamberlain's company whose services were no longer required memorially reconstructed the script for a provincial touring company he had joined. In this he may have been aided by a nonacting accomplice. In order to suit the script to local audiences, this traitor-actor—who had played the Host—cut away all extraneous speeches, excised all references to the Order of the Garter and Court, and made certain alterations in the text based on his playing knowledge. What remained was a swiftly moving comic plot. He then sold the manuscript to Busby who entered it in the Stationers' Register in January 1602, and transferred it to Arthur Johnson for publication. Prevented by the copyright regulations of the Stationers Company from issuing a legitimate, uncorrupted text of the play unless they were willing to permit Busby or one of his associates to publish it (which after Busby's circumventing of their blocking entry in the Stationers' Register for *Henry V* they apparently did not desire), the Chamberlain's Men had to accept the piracy. When the scheme for the First Folio was worked out, Heminges and Condell together with the publishers decided to have Ralph Crane make fresh literary transcripts of Shakespeare's plays. Somehow this plan broke down after *Measure for Measure*, the fourth play in Folio order, had been treated in such fashion. But the *Merry Wives*, as the third play, did get itself perpetuated with the characteristics of act and scene division, character listings grouped at scene heads, an almost complete absence of stage directions, and some oddities in punctuation. Whether it was the author's manuscript or the prompt book which stood behind this literary transcript may never be known. Editors early recognized the superiority of the Folio text, but disturbed by the contamination in it, found

that they had to rely on the Quarto to explain certain readings. The blending of the two texts has resulted in the
creation of an "authoritative" edition. Though this edition
is a synthetic re-creation—the product of the study rather
than of the rehearsal hall—it brings us closer to the script
of the 1597 production than either the Folio or Quarto
texts alone can.

CHAPTER VI ✦ THE BROOKE-BROOME VARIANT

HEN he decides to test the veracity of Pistol's report of Falstaff's plan to woo his wife, Master Ford tells the Host, ". . . Ile giue you a pottle of burn'd sacke, to giue me recourse to him [Falstaff], and tell him my name is *Broome*: onely for a iest" (II.1.222-224). In relating this same incident, the Quarto (lines 432-433) records the name as *Brooke*. In the Folio text when Ford arrives at the inn, Bardolph announces him:

> *Bar.* Sir *Iohn*, there's one Master *Broome* below would
> faine speake with you, and be acquainted with
> you; and hath sent your worship a mornings
> draught of Sacke.
> *Fal.* *Broome* is his name?
> *Bar.* I Sir.
> *Fal.* Call him in: such *Broomes* are welcome to mee, that
> ore'flowes [*sic*] such liquor. . . .
> $\qquad$ (II.ii.150-156)

Here the Quarto reads:
> *Bar.* Sir heer's a Gentleman,
> One M. *Brooke*, would speak with you,
> He hath sent you a cup of sacke.
> *Fal.* M. *Brooke*, hees welcome: Bid him come vp,
> Such *Brookes* are alwaies welcome to me:
> A *Iack*, will thy old bodie yet hold out?
> $\qquad$ (Q. 539-544)

Throughout the Folio text Ford's alias appears as *Broome*. Yet it is apparent from the word play "such *Broomes* are welcome to mee, that ore'flowes such liquor" that the line

[107]

is meaningless unless the Quarto rendering of the name as *Brooke* be substituted. Alexander Pope was the first editor to recognize that the Quarto supplied the correct name although omitting the key qualifying clause "that ore'-flowes such liquor." He accepted the validity of the Quarto version and restored the *Brooke* reading to the Folio text in his 1725 edition of Shakespeare. All subsequent editors have followed him in preparing their copy for the *Merry Wives*.

What remains a tantalizing puzzle, however, is why there should have been a name change. *Broome* appears entirely too many times in the F text to be charged off as a scribal or compositorial error. Few scholars have been tempted to penetrate the problem, but those who have, have produced some intriguing theories.

In approaching the question, Professor John Crofts[1] advanced the argument that the original reading was *Brooke* but that the change was made by William Jaggard during the printing of the Folio in order to side-step a possible controversy with Ralph Brooke, the York Herald. In 1619, Jaggard had published Brooke's *A Catalogue and Succession of the Kings, Princes, Dukes, Marquesses, Earls and Viscounts of this Realme of England.* . . . The book contained many errors which Brooke tried to attribute to Jaggard's printing. As a result, a feud lasting through 1622 broke out between the two men. Brooke brought out a second edition in 1621, printed this time by William Stanby, in which he scored Jaggard for the mistakes. In 1622, Augustine Vincent prepared a violent criticism of Brooke's book which Jaggard published under the title *A Discouerie of Errours in the first Edition of the Catalogue of Nobility published by Raphe Brooke.* . . . With the appearance of this volume, Brooke's work was completely discredited.

[1] *Shakespeare and the Post Horses*, pp. 103-105.

In postulating Ralph Brooke as the culprit behind the Brooke-Broome name change in the *Merry Wives*, Crofts completely misestimates both the power of Brooke and the relationship between Brooke and Jaggard.[2] Brooke was very unpopular with his colleagues at the College of Arms and had even become involved in bitter feuds with various of them. Nor did his position as a herald carry sufficient weight to make Jaggard fear him. In fact, Jaggard made his contempt for Brooke public by printing a defense of his printing skill as a preface to Vincent's book.[3] If further evidence were needed to expose the tenuousness of Crofts' theory, it would be found in David White's very practical observation in "The Textual History of 'The Merry Wives of Windsor'" (p. 103) that Jaggard merely had been handling Shakespeare's material—printing the lines Heminges and Condell had furnished him with. And Brooke could not hold the printer responsible for that.

[2] For the most complete account of the Brooke-Jaggard-Vincent quarrel as well as a study of Brooke's relationship with other members of the College of Arms, see Nicholas Harris Nicolas, *Memoir of Augustine Vincent, Windsor Herald* (London, 1827), pp. 21-72. See also Edwin Eliot Willoughby, *A Printer of Shakespeare: The Books and Times of William Jaggard* (London, 1934), pp. 144-156.

[3] Jaggard sums up his case stating, "I hope I have . . . washt my hands cleane of that aspersion which he casts upon me concerning the Errors of his Booke, the main Error excepted of defiling my fingers at all with his pitch which cleaves so fast to my hands as I shall never shake them off but with losse; wherein if ever againe I be taken faulty, let that curse light upon me, which I prophesie will befall any Printer that hath to doe with him: that he worke by day in feare and like a theefe in the night: that he bring forth his works like bastard fruits by stealth and vent them in corners like stolne Wares, and that in recompence of all his paines his reward be no other but losse and repentance by the Worke and detraction and disgrace from the author. . . . I have satisfied myself . . . that M. Yorke [York Herald] may understand, it touches a Printer as much to maintaine his reputation in the Art he lives by, as a Herald in his Profession, and that if any affront be done me in that kinde, I shall be ever as sensible of it, as hee would be of the like done to himselfe: howsoever it hath pleased God to make me, and him to style me a Blinde-Printer, though I could tell him by the way, that it is no right conclusion in schooles, that because *Homer* was Blinde and a *Poet*, therefore hee was a *Blinde-Poet*. FAREWELL." Quoted by Willoughby, pp. 155-156. The two ellipses within sentences are Willoughby's.

White himself has postulated another theory.[4] Shortly after James had ascended the throne, the Court was rocked by the Bye Plot. One of the chief supporters of William Watson in this Catholic conspiracy against the monarch was George Brooke, brother of Henry, Lord Cobham. On December 5, 1603, Brooke was beheaded for his role. At the same time that the Bye Plot was broken, another conspiracy, the Main Plot—designed to place Arabella Stuart on the throne—also was detected. Lord Cobham, one of the instigators of this plot, was convicted of high treason. Though sentenced to death on December 10, he received a last-minute reprieve from the King and spent most of his remaining years as a prisoner in the Tower. Almost a year after the breaking of these plots, on November 4, 1604, the King's Men gave a Court performance of the *Merry Wives* which James attended. In preparation for this performance, White believes that the King's Men altered the name *Brooke*, which was the original one, in order to avoid irritating James by arousing unpleasant associations which that name might have for him as a result of the role of the Brooke family in the Bye and Main conspiracies.

Interesting as this theory is, it is predicated too much on the sensitivity of James to a name which was fairly common throughout England. Since there is nothing in the *Merry Wives* to mirror events of either the Bye or Main plots, it seems rather far-fetched to believe that the King would have become rankled at the name *Brooke* almost a year after George Brooke had been beheaded. If the name

---

[4] "Textual History," pp. 103-105. This is the same theory which Alfred Hart developed simultaneously in *Stolne and Surreptitious Copies*, pp. 89-90. Apparently unaware of these prior investigations, William Bracy reiterates the argument in *History and Transmission*, p. 135. I arbitrarily attribute the theory to White because it receives the most extensive treatment by him. See also Samuel R. Gardiner, *History of England from the Accession of James I. to the Outbreak of the Civil War 1603-1642*, new impression (London, 1900), I, 108-140.

had become an anathema to him, why not predicate a mass banishment of all Brookes from Court? If so, Ralph Brooke, the York Herald, might not have become embroiled with Jaggard.

Not even a suggestion of prudence or forethought in altering the name to *Broome* can be considered as tenable, for there are ample contemporary documents attesting that the King did not manifest deep hostility toward Lord Cobham, the surviving Brooke brother. And his attitude was common knowledge. Indeed, a scant five days after the execution of George Brooke, Dudley Carleton is found writing to John Chamberlain that the King believed Cobham "had shewed great tokens of humility and repentance"; and when the King announced he would not execute Cobham, Grey, and Markham, "the applause that began about the king went from thence into the presence [presence chamber], and so round about the court."[5] Similar commentary appears in a note of Sir William Browne to Lord Sydney on December 31, 1603, in which Browne states, "The King's mercy towards the offenders, Cobham, Grey and Markham is much admired."[6] Further evidence comes from another source: the manner in which the expulsion of Henry, Lord Cobham from the Order of the Garter was handled. Since he had been convicted of treason, Cobham "had befouled the glory of that most noble order." Accordingly, on February 7, 1604, the Garter King of Arms received a warrant to publish Cobham's degradation (*Cal. S.P. Dom., James I, 1603-10*, VI, 53). On February 12, the ceremony was carried out at Windsor. Standard procedure in such cases called for the Garter King of Arms to throw down violently the knight's achievements which hung above his stall, to kick them out of St. George's

[5] Thomas Birch, ed., *The Court and Times of James the First...* (London, 1849), I, 32.
[6] H.M.C. *Penshurst*, III, 82.

Chapel, and finally to cast them into the ditch outside the castle. But there is a marked departure in this particular degradation ceremony. After the King of Arms had thrown down and, in the presence of many onlookers, stamped on Cobham's achievements, he kicked them only as far as the door of the Chapel and no further; for James, out of regard for Cobham's nobility, had expressly forbidden the achievements to be cast into the common ditch.[7] Such deliberate action by the King in breaking Garter procedure can only be construed as an additional indication that James would not have flown into any rage at the mention of the name *Brooke*. Thus any hypothesis advocating an alteration to the *Broome* reading during the preparation for the Court performance of the *Merry Wives* in November 1604, must, like the achievements of a degraded Knight of the Garter, be cast out.

In contrast to the theories of Crofts and White which argue for the name alteration at a later date, stands the intimation of Leslie Hotson that the change may have been made for the original 1597 production.[8] Hotson suggests that the jealous Ford received the alias of Brooke deliberately as an intentional affront to William Brooke, Lord Cobham, whom the players had reason to dislike at this time. Their hostile feelings, Hotson believes, stem from the growing opposition of the Lord Mayor and Aldermen of London to the operation of the playhouses. With the death of Henry, Lord Hunsdon in July 1596, the players had lost their chief protector in the Privy Council. His successor, Lord Cobham, "did not," Hotson maintains, "look with favour on the players."[9] Thus Shakespeare re-

---

[7] *Blue Book*, p. 153. For comparison with the usual procedure in such cases, see the description of the degradation of Thomas, Earl of Northumberland on November 27, 1569, in B.M. Harl. MS. 6064, fol. 39v; also that of James, Duke of Monmouth on June 16, 1685, in B.M. Harl. MS. 6834, fol. 3. See also Ashmole, p. 622.

[8] *Shakespeare versus Shallow*, pp. 13-15.

[9] *ibid.*, p. 15.

taliated, both in the *Merry Wives* and in *Henry IV*, where he originally labeled the fat knight *Oldcastle* after Cobham's distinguished ancestor. But when offense was taken, the names were altered to *Falstaff* and *Broome*.

While Hotson presents this theory with great reservation, in doing so he only adds to a cloudy legend with which scholars have enveloped Lord Cobham. It is true that ever since the 1570's there had been a sniping war between the London city officials and the theater people and that between 1592 and 1596 the city officials intensified their efforts to suppress the players. But the charge that Lord Cobham had Puritan leanings and accordingly used his influence to force the passage of antitheatrical regulations does not seem based on any concrete evidence. If Cobham did follow Puritan doctrines, contemporary documents apparently make no mention of it. Therefore any statement that because his ancestor Sir John Oldcastle was a Lollard, Lord Cobham must have subscribed to the Puritan way of thinking has to be considered as mere allegation.

Cobham served as Lord Chamberlain from August 8, 1596, until his death on March 5, 1597. During this period not one piece of legislation hostile to the theatrical interests was enacted. In fact, the first decree for *permanent* suppression of a public theater was issued on July 28, 1597, four months after Cobham's death. But between 1597 and 1603—during the tenure of George, Lord Hunsdon, as Lord Chamberlain—attempts to restrain the players were numerous and effective. Moreover, while it is impossible to surmise what behind-the-scenes maneuvering may have taken place, the Privy Council records show that Cobham attended few meetings of the Council between 1592 and 1596. More startling, in light of Cobham's supposed hostility toward the theater is the fact that during this period Lord Cobham was absent from every meeting of the Council at which a restraining piece of theatrical legislation was passed.[10] Oddly enough, there is further ground to suspect that Cobham would not have gone out of his way to be antagonistic toward the players. John Tucker Murray in *English Dramatic Companies* makes reference to a troupe known as Lord Cobham's players operating in 1563 and to the Lord Warden of the Cinque Ports Company, which had Cobham as its patron. The records are few and far between for the years 1563-1571; yet, one wonders, unless a radical change came over Cobham, would he by the 1590's have become so antitheatrical?

Since the persecution of the players was more intense during the time both Lords Hunsdon were controlling forces in the government, it seems more reasonable to conclude that the Hunsdons, with their keen interest in the drama, made special efforts to protect their players. But Lord Cobham simply may not have manifested as much concern with this particular problem. If this were the case, Cobham was not responsible for repressing the dramatic companies, and he has been much maligned by writers who, in erroneously equating indifference with enmity, have created their own special brand of scholarly folklore. Therefore, to propound an intentional affront to Cobham in Shakespeare's employing the names *Oldcastle* and *Brooke* is to build a theory without a foundation.

The choice of the name *Oldcastle* in *Henry IV* can more properly be explained by the fact that Shakespeare merely took over the character of Sir John Oldcastle from *The Famous Victories of Henry the Fifth*, never realizing that he might cause some wincing among the descendants of the real Oldcastle. When the complaint was made, Shakespeare quickly altered the name to *Falstaff*. While there is general critical acceptance of the theory that *Oldcastle* was

[10] June 23, 1592, January 28, 1593, July 29, 1593, July 22, 1596. Since the Privy Council records for 1594 are not extant, the decree of February 3, 1594, cannot be commented on.

changed to *Falstaff* as a result of protests from the Brooke family, a divergence of opinion exists as to who instituted the complaint. Chambers and Greg believe it was Henry, Lord Cobham, whereas Dover Wilson contends it was his father William.[11] The crux of the problem lies in the dating of *1 Henry IV*. Those who adopt the more conservative date of late 1597 naturally attribute the objection to Henry, Lord Cobham since his father had died the previous March. But (as will be shown in Chapter IX), there is strong reason for dating *1 Henry IV* between July 1596 and March 1597. In that case there can be little question that William, Lord Cobham forced the change. Actually, the elder Lord Cobham was more in a position to demand immediate remedial steps for this accidental slur on his family name. As Lord Chamberlain he had the power in Court as well as the authority to order either directly or through the Master of the Revels that the alteration be made.

The renaming of *Oldcastle* as *Falstaff* has direct bearing on the manner by which I believe *Brooke* became *Broome* in the Folio copy of the *Merry Wives*. Shakespeare selected the name *Brooke* without any thought of maligning William Brooke, Lord Cobham. Rather he was motivated in his choice by the very nature of his material. In the scene in which Pistol and Nym relate Falstaff's plot to Ford and Page (II.i.), Ford becomes tremendously disturbed and resolves to take some form of action, inquiring of Page whether Falstaff "lye at the Garter." At this point the Host of the Garter and Shallow arrive and invite Ford and Page

---

[11] Chambers, *William Shakespeare*, I, 383. But earlier, in *The Elizabethan Stage*, II, 196, Chambers states, "It is not improbable that the offence taken was by Lord Chamberlain Cobham, whose ancestress, Joan Lady Cobham, Oldcastle had married." Greg, Review of *The Merry Wives of Windsor, The History and Transmission of Shakespeare's Text*, by William Bracy, SQ, IV (January 1953), 79, and *The Shakespeare First Folio*, p. 262, n.5. J. Dover Wilson, "The Origins and Development of Shakespeare's *Henry IV*," *Lib.*, 4th Ser., XXVI (June 1945), 12-16.

to join them in their little sport with Evans and Caius. It is implicit from the text that Ford has been improvising some plan which at the sight of the Host he immediately resolves to embark upon. He will meet with Falstaff in disguise and under an alias. What name should he pick? Pressed into making a quick decision, he chooses something associated with Ford. Ah yes——Brooke.[12] Even the most cursory reading of the *Merry Wives* reveals that the dialogue abounds in a variety of types of word play—puns, malapropisms, etc. Making *Brooke* the alias for *Ford* seems to be just the turn of mind Shakespeare was in while writing the play. Once having established this water image, Shakespeare plays on it twice more in the script. Thus the conceit "Such Brookes are welcome to me, that ore'flowes such liquor." And in III.v. after Falstaff returns to the inn from his unexpected bath in the Thames, Mistress Quickly arrives to entice him to make a second visit to Mistress Ford:

> *Qui.* Marry Sir, I come to your worship from M. *Ford.*
> *Fal. Mist. Ford?* I haue had Ford enough; I was thrown into the Ford; I haue my belly full of Ford.
>
> (III.v.34-38)

Modern editors have emended all the upper case *f's* to lower case in Falstaff's line. But if we may assume the accuracy of the original Folio reading, we recapture exactly what Shakespeare was doing with the names *Brooke* and *Ford.* It is therefore proposed that the foul papers carried the reading *Brooke,* and in this form the play went into rehearsal for the 1597 Garter performance.

Now somewhere in the course of rehearsing for that performance someone realized that Brooke was also the name

---

12 I am indebted to Dr. Randolph Goodman of the Department of English, Brooklyn College, for noting the association between the two names.

of the late Lord Cobham who had objected so strenuously to the appearance of Oldcastle in *1 Henry IV*. Aside from the fact that the character of Ford is not exactly an engaging one—that of a jealous husband who would deliberately hire someone to cuckold him—once the discovery of the coincidence in names had been made, Shakespeare's company probably decided to show caution and avoid any repetition of the unpleasantness which arose over the Oldcastle-Falstaff affair.[13]

There were imperative reasons for the company to exercise restraint in this really trifling matter of "what's in a name." Yes, Lord Cobham was undoubtedly the individual who had complained about the blackening of his family name in a Shakespeare play only a few months before the *Merry Wives* production. And Cobham had been an important man at Court—a member of the Privy Council since about 1586, also Lord Chamberlain. Only a few months before he received the latter post, the Queen had even remarked in a letter to him (June 7, 1596), "We have long had proof of Your faithful service, as Lieutenant of Kent, and Warden of the Cinque Ports. . . ."[14] Cobham had died less than two months before the *Merry Wives* production and was interred at his estate in Kent on April 5, 1597. Thus it would have been foolhardy for Shakespeare's company to have risked doing anything which could have been misconstrued as an insult to the memory of this powerful lord.

A further possible association with Lord Cobham produces yet another deterrent to the use of the name *Brooke* at this time. The first performance of the *Merry Wives* was,

[13] A somewhat similar thought had earlier occurred to Frederick Gard Fleay who states in *A Chronicle History of the Life and Work of William Shakespeare. . .* (London, 1886) that the name "was probably altered because Brooke was the name of Lord Cobham, who took offense at the production of Oldcastle on the stage" (pp. 210-211).

[14] *Cal. S.P. Dom., Eliz., 1595-1597*, p. 226.

in my opinion, presented as a salute to the Order of the Garter. Lord Cobham, who had been elected to the Order in 1584, had been an influential member of that body. In 1595, the Queen had appointed him to serve as her Lieutenant for the celebration of that year's Feast of St. George. Commenting on such an appointment, Ashmole indicates, "As to the Person Nominated, we observe, That (most usually) he hath been one, if not the chief of the Knights-Companions (we mean in Authority, Eminence, or Birth) next to the Sovereign himself. . . . The Dignity of this Officer, as he represents the Sovereign's person, and supplies his place, is very great."[15] So high an honor was it to serve as Lieutenant in the sovereign's absence that the individual thus appointed was excused from attending the following three years' Feasts if he so desired.[16] Cobham was still basking in the afterglow of his lieutenancy, then, when death closed his career. Even from the grave he could have waved a restraining hand at the accidental use of *Brooke* in the *Merry Wives*.[17]

Which member of Shakespeare's company became aware of the danger of using the name *Brooke* in the *Merry Wives* it is useless to conjecture. It may even have been Lord Hunsdon who, in his dual capacity as Lord Chamberlain and master of the company, decided to watch one of the final rehearsals of the play to make sure it was a fit and proper production for the occasion. Hunsdon would have been especially sensitive to any allusions which might re-

[15] Ashmole, pp. 530, 536. Also see above, Chapter II, Breuning's description of the 1595 Feast and the citation of Cobham to the lieutenancy.
[16] Ashmole, pp. 537-538.
[17] On this point, Ashmole (p. 624) observes: "for as the reputation which the Knights-Companions, while living, derived from their admission into so renowned and illustrious a Body, specially Knights-Subjects, who were thereby advanced to a fellowship with their King and Supreme Lord, and made Companions to Emperors, Kings, and Princes, was very great; so were the several Honors paid to their memory after their decease. . . ."

flect detrimentally on a Knight of the Garter. At any rate, the *Brooke* error was detected, and a last-minute change to *Broome* made.

There is good reason to suspect a hasty job of altering. The substitution of an *m* for a *k* in the players' parts and the prompt book it easily accomplished. But surely Shakespeare would have realized he had lost his pun with the revised reading "such *Broomes* are welcome to mee, that ore'-flowes such liquor. . . ." At the last minute he probably found it too complicated to rewrite the whole speech, and knew that in performance the botched pun would pass unnoticed. Had he not had a similar experience in *1 Henry IV* when he altered *Oldcastle* to *Falstaff*? There was that twist to "my old Lad of the Castle" in I.ii.47-48 which he retained even though the reference was now meaningless. Again he let a botched pun go through; and because of lack of interest in accuracy in printed versions of plays, it remained in the script right down through the printing of the Folio.

But how did the Quarto get the original reading *Brooke*? Let us remember that the individual held instrumental in making the memorial reconstruction of the *Merry Wives* is the actor who played the Host. The Host is the first character who learns that Ford has adopted the alias of Brooke, and he promises accordingly "thy name shall be *Broome* [Brooke]." Thus, the Host would have been one of the actors called upon to relearn the line. He would also have been aware of the efforts of the other actors to memorize the new name at the last minute. (It appears forty-two times in the Folio.) And he possibly knew the reason behind the change. When he came to reconstruct the play, he remembered the incident and restored the *Brooke* reading to the text.

The Quarto shows other instances in which inconsistencies in the Folio version have been resolved through what can only be interpreted as solutions arrived at by an actor who had actually rehearsed and performed in the *Merry Wives* with the Lord Chamberlain's Men. The problem of a distinctive color for Anne's costume in V.v. has been eliminated in the Quarto. Also, this text has corrected the confusion in names when Pistol and Nym decide to betray Falstaff to Ford and Page.[18] These changes augur the work of a memorial reconstructor intent on making a tight script as a result of his playing knowledge.

This individual no longer had to fear any possible affront being taken at the name *Brooke*. With the reconstruction made for a provincial tour, the chances of anyone witnessing a performance who might object to a *Brooke* among the characters are slim. Furthermore, with the excision of all dialogue in the Quarto which touches on Court affairs, what remains is a play cast in a rustic setting. Thus *Brooke* becomes simply a common Elizabethan name and settles among the Fords and Pages without attracting any undue attention to itself.

So the pirated Quarto retained the original reading of *Brooke* while the mother text of the Folio carried the revised *Broome* entry until Alexander Pope set things aright once again. Insignificant as the point may be against the larger field of textual alteration, the *Brooke-Broome* variant has behind it, in my opinion, an intensely human drama—one that could have been enacted, in all probability, only at the time of the composition of the *Merry Wives* in 1597.[19]

[18] In the F text, Nym is to inform Ford and Pistol, Page (I.iii.104-105). But in II.i.113-141 Pistol addresses Ford and Nym, Page. In the Quarto in both instances (lines 228-231 and 361-375) Pistol carries the news to Ford, and Nym speaks with Page.

[19] My research has yielded no individual named Broome who might have come into sudden prominence to warrant a topical allusion to him at a Garter feast play. Hence the discussion in this chapter precludes a consideration of a deliberate change from *Brooke* to *Broome* to reflect on someone bearing that name.

CHAPTER VII ✦ THE "DUKE DE JARMANY"

THE horse-stealing subplot of the *Merry Wives* revolves around the arrival of a German duke at Windsor. In his name, his retainers secure three horses from the Host of the Garter Inn and then run off with them. Immediately after Bardolph reports the loss (IV.v.64-74), Evans enters warning the Host (in the Quarto version) that "there is three sorts of cosen garmombles, is cosen all the Host of Maidenhead & Readings." Dr. Caius follows with the pronouncement that "it is tell-a-me, dat you make grand preparation for a Duke *de Iamanie*: by my trot: der is no Duke that the Court is know, to come. . . ."

Is this brief allusion to a "Duke de Jarmany" solely an invention of Shakespeare's to further one of the subplots of the *Merry Wives*—a subplot which went down in the wreck of hasty composition? Or did the dramatist have in mind some contemporary event and an actual duke whom he wished to portray? Charles Knight, in 1839, called attention to the fact that a German duke, Frederick of Württemberg, had indeed visited Windsor in 1592 and that "the circumstance would be one of those local and temporary allusions which Shakespeare seized upon to arrest the attention of his audience."[1] What attracted Knight to the visit of this particular German nobleman was that it had occurred in 1592, the year before what Knight believed was the first sketch of the *Merry Wives* had been completed. Though he erred in his reason for selecting the

[1] *The Pictorial Edition of the Works of Shakspere*, "Comedies," I, 143. See also p. 144. For further commentary on the identification of the "Duke de Jarmany" with Frederick, see Rye, *England as Seen by Foreigners*, pp. xciv-ciii; Arden *Merry Wives*, pp. xli-xlvi; New Cambridge *Merry Wives*, pp. xix-xxi; Chambers, *William Shakespeare*, I, 427-429.

Duke of Württemberg—especially as Frederick at that time had not succeeded to the dukedom and was still only Count Mompelgard—Knight did identify the sole possible model for the Duke de Jarmany. For this same Duke was elected to the Order of the Garter in 1597, and the allusion in the play probably has its roots in this Garter link.

But why should Shakespeare have taken such pains to build a topical allusion around the Duke of Württemberg? What was so unusual about this man? What did he mean to Queen Elizabeth and her Court? Historically, Duke Frederick has not transcended his time; but for a brief period of twelve years, from 1592 to 1604, he strutted and fretted upon the stage of Anglo-German relations. Hence, before we turn to the horse-stealing episode itself, it is fitting that we pause to make the acquaintance of Frederick, Duke of Württemberg and Teck, Count Mompelgard.

Count Mompelgard arrived in England on August 9, 1592, and spent a month traveling through the countryside. From August 17 to 19 he was at Reading where the Queen granted him two audiences and where the Earl of Essex and other nobles feasted him. He then went to Windsor for a few days where he hunted and went sightseeing. On the 22nd he was in London and on the 25th commenced a visit to Oxford and Cambridge. He returned to London on September 1 and left for Württemberg on the 4th. Before departing he received the gift of a horse from Essex. The following year, on August 8, he succeeded to the title of Duke of Württemberg.[2]

As a ruler, Frederick seems to have governed well dur-

---

[2] An account of his journey was made by his secretary Jacob Rathgeb and published under the title *Kurtze und Warhafte Beschreibung der Badenfahrt. . .* (Tübingen, 1602). It was translated by W. B. Rye in *England as Seen by Foreigners*, pp. 1-53, from which source the above information has been digested. All references from the *Badenfahrt* in this study are cited under Rye. Knight had come across a copy of the German original about twenty-five years before Rye's translation appeared.

ing the fifteen years of his reign.[3] He initiated many projects for the betterment of his state, stimulated business and trade, supported the arts and sciences, even succeeded in reducing the public debt. In international affairs he frequently served as a peacemaker. An ardent Protestant, he fought against the spread of Catholicism by building alliances with the Protestant princes of Germany—a point which would have made him particularly useful to Queen Elizabeth.

In his personal life, however, he appears to have acted in an entirely different manner. Here extravagance, vanity, and ostentation ruled him. And these led him into a consuming passion to become a Knight of the Garter. Membership in the Order became an obsession with him, driving him to write letter after letter and send ambassador after ambassador to England. He pleaded with, wheedled, and cajoled Elizabeth and her ranking courtiers to admit him to the Order. In this—and in this matter only—did Frederick achieve notoriety in the English Court.[4]

One can well imagine then the effect on him when he attended a service in St. George's Chapel during his visit to Windsor that August in 1592 and observed hanging "in the said church . . . the shields, helmets, and banners of the knights of the royal order called the Garter (*La Chartiere*), which [as his secretary Rathgeb notes] is a highly esteemed order, and which not many can obtain."[5] As Frederick gazed at the magnificent regalia and later dined

---

[3] He died January 29, 1608. The biographical information on Frederick has been gathered from: "Friedrick I., Herzog von Würtemberg," by P. Stälin, *Allgemeine Deutsche Biographie*, VIII, 45-48; S. L. Millard Rosenberg, "The Original of the 'Duke de Jarmany,'" *University of California Chronicle*, XXXV (January 1933), 90-93; Rye, pp. lxxxvii-lxxxviii; Klarwill, pp. 347-355.

[4] There have been attempts to link Frederick with English post-horse scandals. Professor Crofts, through his study of contemporary documents, has found absolutely no evidence to support such a contention (*Shakespeare and the Post Horses*, pp. 11-18).

[5] Rye, p. 16.

with the attendant Poor Knights, his mind undoubtedly went back to his conversation with Queen Elizabeth only a few days before in which he requested her to make him one of that noble Garter band.

Rathgeb makes no mention of the discussion, but it is clear from a letter the Count wrote to Elizabeth on March 23/April 2, 1593, that the matter had been broached: "Your Majesty will doubtless remember what *I in my own person humbly asked of you, together with the favorable reply made to me.* With this object in view, and because the proper time is near at hand [the Feast of St. George], I have despatched this bearer, a gentleman and good soldier, to solicit my affairs, trusting to receive by him a favorable and much wished-for answer.[6] This was the first of the letters rained on Elizabeth and important nobles over the next decade.[7]

The Queen ignored Frederick's request when the knights met in chapter on St. George's Day, 1593, to elect their new comrades. Though it lay entirely within her province as sovereign to direct the election of any foreign prince to the Order, Elizabeth may have felt that a nobleman not a head of state was too minor a foreign functionary to be considered for election.

But Count Mompelgard wanted that Garter. In fact, so obsessed had he become with receiving it that shortly after he became the Duke of Württemberg in August 1593, his

---

[6] P.R.O. SP 81/7, fol. 121. I have used the translation which appears in Rye, p. lxi. Italics Rye's. I am indebted to Rye's listing of the locations of the extant correspondence between Württemberg and England. There have been relatively few additions or subtractions in the collections since Rye examined the documents in the 1860's.

[7] Mention of a promise of Queen Elizabeth to make Frederick a Knight of the Garter also appears in Erhardus Cellius, *Eques Auratus Anglo-Wirtembergicus* (Tübingen, 1605), pp. 73, 78 (marginal note). Inasmuch as this book was written to commemorate the investiture of Frederick into the Order in 1604, it is composed in a highly laudatory prose style. Though it relates the entire story of Frederick and the Garter, the earlier portions are undoubtedly based on Rathgeb's *Badenfahrt*.

vanity got the better of him, and he had coins struck portraying him wearing the Collar of the Order; on the reverse were his arms encased within the motto.[8] And the following February he sent Elizabeth a reminder of his request (P.R.O. SP 81/7, fol. 183). However, 1594 was a year without a Garter election. The Queen merely replied on May 17 to Frederick's latest communique with a polite note (P.R.O. SP 81/7, fol. 194) acknowledging her promise but temporarily putting off any action.

Having failed in his first two attempts to get the Garter, Frederick tried a new tactic as the 1595 Feast of St. George approached. This time he sent Hans Jacob Breuning von Buchenbach, one of his most skilled diplomats, to London to intercede for him. Breuning, who was assisted in this mission by another reliable envoy, Benjamin Buwinckhausen (or Bouwinghausen) von Wallmerode, arrived in England in late March. Breuning wrote a complete account of his mission entitled "My most submissive narration of all that happened from the day on which I was graciously dispatched to England by the Court till my return, and of all that I have with the utmost sedulousness accomplished."[9] This document is most important for giving us a direct look at the machinations Breuning became involved in on behalf of his master, as well as for providing some insight into the impression Frederick had made upon the English court and some indication of how fixed an impression it was.

On April 2, Breuning saw the Earl of Essex (who had been so friendly to Frederick during the latter's 1592 visit)

---

[8] Klarwill, p. 353. See illustration opposite p. 394, that book.

[9] This was published under the title *Relation über seine Sendung nach England im Jahr 1595* by the Bibliothek des Litterarischen Vereins (Stuttgart, 1865). Another copy prepared from the MS in the Württemberg archives was included by Victor von Klarwill in *Queen Elizabeth and Some Foreigners*, pp. 357-423. Klarwill has made a few very minor omissions. I use the transcription of Klarwill in the Nash translation. For the biographical background on Breuning, see Klarwill, pp. 351-352.

and asked his help in arranging an audience with the Queen. Breuning also paid a visit to Lord Burghley on the same matter. Both noblemen received him cordially, with the result that Breuning obtained his audience for April 6.

On that day, as Breuning relates, when he came before the Queen, "Both the Privy Chamber and the Presence Chamber were full of Mylords, Grandees, Earls, Lords, and of very grand countesses and Ladies, who were all without exception beautiful."[10] Before this crowd he turned to Her Majesty, addressing her in Italian—"which language I had been told, would sound sweetest to her ears," and pleaded with her to honor her promise made three years earlier to receive Frederick into the Order of the Garter.[11] The Queen told him that she would give her answer at a later audience. In the meantime, she asked him to prepare a written copy of the address he had just delivered.

While waiting for that audience Breuning further pressed his campaign. On April 8, he closeted himself with Lord Burghley, delivering a letter from Frederick and openly asking for Burghley's intervention in getting the Duke elected to the Order.[12] On the ninth Breuning wrote to Burghley (B.M. Lansd. MS. 79, fol. 64v) repeating his request of the previous day and freely acknowledging that the primary purpose of his trip was to acquire the Garter for Frederick. This was but one of a series of letters which he wrote to various people at Court on this subject.

Not content with mere correspondence, Breuning planned greater intrigues to accomplish his ends.[13] Accordingly, he "was at the greatest pains to get knowledge of a person in a lower station of life, who might have the ear of Her Majesty." He succeeded in finding a German jeweler from Lindau by the name of Johannes Spielmann who was

[10] Klarwill, p. 363.   [11] ibid., pp. 364-365.
[12] ibid., pp. 366-367.
[13] These are recounted in Klarwill, pp. 369-370.

in favor with the Queen as well as with Lord Burghley and Sir Robert Cecil. Not satisfied with one go-between, Breuning took further precautions as the time for his second audience with the Queen drew near and attempted to bribe a M. de la Fontaine with four hundred crowns, Spielmann with four hundred, and the secretary to the Earl of Essex with one hundred "if the affair were brought to a successful issue." Although the bribes were refused "more from a sense of honour than because they were loth to accept," the three men did try to influence Lord Burghley, Sir Robert Cecil, the Earl of Essex, and even "the son of My Lord Cobham" to act in Frederick's behalf.

Apparently Breuning was having some success with his lobbying, for he continues his report: "Having further learned that all the gentlemen, and not least the Lord High Treasurer, were practically won over to our side, I could give credence to the intelligence that the only obstacle to my affair was that the Garter or insignia of the Order had not yet been despatched to the King of France and the King of Scotland, and that, therefore, it would this time be almost impossible to elect Your Grace or any other into the Order. These tidings were imparted to me not only by M. de la Fontaine and Spielmann, for the matter formed a topic of discussion among the Knights of the Order; but the Earl of Essex himself and Sir E. Stafford intimated as much to me. I therefore deemed it necessary to dissipate such doubts before the advent of St. George's Day, and accordingly, on the Saturday before Easter, wrote both to the Earl of Essex and to the Lord High Treasurer. . . ."[14]

St. George's Day arrived with Breuning much in doubt about what would happen. He received an invitation to

14 *ibid.*, p. 371. An interesting marginal note has been appended in the manuscript by Duke Frederick who comments on the passage about the impossibility of electing the Duke as "always the same old story." Breuning gives the text of his letter to Essex, which the Earl never answered, on pp. 371-372.

attend the ceremonies, and, after due deliberation decided to accept. After the Grand Dinner Breuning broached the subject of the Duke's membership to three of the Knights of the Garter. Their reply was "that the choice of foreign Knights lay principally with Her Royal Majesty; but as concerned themselves they would do their best, as they felt themselves to be under obligation to serve Your Princely Grace."[15]

But the Queen and the Knights-Companions had jointly decided not to hold any election in 1595, although deaths since the last election in 1593 had brought the total of vacant stalls to five. We can well imagine, then, the disappointment of Breuning as the Feast of St. George ended with the naming of no new members.

Two days after its conclusion, on April 26, Breuning had his second audience with Elizabeth. This time few persons were present and none within hearing. The Queen asked the ambassador to speak frankly, and he briefly repeated his earlier request of obtaining the Garter for Frederick. To this the Queen replied that although favorably disposed to the Duke, she could not appoint him at that time because "some Kings were a long time past elected by the corporate consent of the entire Order and associated with it, and that to these Kings the insignia have not yet been remitted. Accordingly it is absolutely necessary that this be done prior to aught else, and that meanwhile no other be elected."[16]

Then Elizabeth startled Breuning by continuing: "But to speak truth, I have not the least recollection of ever having made any such promise, as I expressly told the then envoy; for by reason of our laws which I have already brought to your notice, I could never have done so without great ignominy to myself and to the detriment of the

[15] *ibid.*, p. 384.   [16] *ibid.*, p. 388.

Kings previously indicated."[17] Though Breuning was shocked at this denial, the Queen assured him that she would do what she could. "But it does not lie with me definitely to promise that I shall confer upon him this honour."[18] Breuning asked Elizabeth for a letter of explanation to the Duke, which the Queen agreed to write.

Before leaving England, Breuning paid a final visit to Essex on May 10, expressing to the Earl his disappointment over Frederick's failure to receive the Garter but hoping that at the first opportunity the Duke would be named to the Order. Essex agreed to help further the matter. Breuning also made final solicitations of Sir Robert Cecil and Lord Burghley. Burghley replied that "no deviation could be made from Her Majesty's resolve" and that it was useless to appeal to the individual Knights of the Garter over her head. On May 18 Breuning sailed for home. In summing up his mission for Frederick, Breuning felt that although the Duke's handling of his attempts to be named to the Order up to that time had largely been bungled, "Your Grace's affair has this time been more forwarded both with Her Royal Majesty and the prominent men than ever before, and is now, thanks to the unremitting soliciting, on the up-grade. . . ."[19]

17 *ibid.*, pp. 388-389.   18 *ibid.*, p. 389.
19 *ibid.*, p. 412. Breuning, in his summation, called attention to the adverse effect the questionable business transactions of the Württemberg merchant Stamler had upon the Duke's Garter campaign. Stamler had come to England to purchase cloth. He had presented false credentials to Burghley that he was acting on behalf of the Duke of Württemberg and involved himself in some shady dealings to export the cloth duty-free. Breuning finally exposed him and put an end to his activities, but not before there had been much court gossip on the matter. The letters which Stamler presented were illicitly obtained through the Duke's secretary. They have been preserved in B.M. Cott. MS. Vesp. F. III, fol. 178; B.M. Lansd. MS. 76, fol. 157; B.M. Lansd. MS. 79, fols. 72-73. The discrepancy between Rye's date of the second letter as December 12, 1593, and its recorded date of December 12, 1594, is explained by the fact that Stamler altered the year of the original to 1594, the date under which it was filed in the archives. The role of Stamler in the Mompelgard story has largely

The letter of explanation for the delay which Elizabeth had sent to the Duke, as Breuning had requested, received a reply on July 4/14, 1595, in which Frederick agreed to wait for the Queen to take action on her promise and begged her not to become annoyed at *"mes sy souuentes solicitations et recharges."*[20] It seems, however, that though Breuning had returned to Württemberg and Frederick had called a momentary lull in his letter-writing campaign, some representative of the Duke had remained in England to serve as a physical reminder of Frederick's desire. On October 15, 1595, in his newsletter to Sir Robert Sidney, Rowland Whyte observes, "The Duke of Wirtemburgs man [Benjamin Buwinckhausen?] is here yet, and like to be long, for he expects a dispatch from the Queen, my Lord Treasorer and others."[21]

The foregoing detailed account of the Breuning mission becomes significant on several grounds:

1) It clearly establishes that great numbers of people both in and out of courtly circles were well aware of the German duke's desire to receive the Order of the Garter. In fact, several had even been offered sums of money for their assistance. And many had heard Breuning openly plead for his prince in that first public audience with the Queen. Thus any allusions to a German duke in a Garter play two years later can neither be considered too esoteric for an audience nor past recollection, especially since Frederick remained a topical figure in the interval.

2) Both the Queen's sudden denial of any recollection of having made a promise to the Duke and her statement that no new foreign rulers could be taken into the Order until the insignia had been delivered to those rulers pre-

---

been misunderstood and magnified by critics. For Breuning's account of the incident, see Klarwill, pp. 359, 398-409.

[20] P.R.O. SP 81/7, fol. 209.
[21] *H.M.C. Penshurst*, II, 172.

viously elected but not yet in receipt of such insignia do
not appear to be borne out by facts. Her letter of May 17,
1594, to Frederick recognizes the existence of such a promise: "As to what you have reminded us of a promise made
of our Order, we pray you to take in good part the reply
which we formerly gave on this subject (you being here)
to the Ambassador of the most Christian King our Brother,
—viz. that seeing there are sovereigns and princes, our
neighbours, accustomed, from time immemorial, to be received into the said order, who are not yet admitted,—
even those who although elected some years ago have not
obtained investiture, we could not incur the remark of a
remissness towards them, and some other princes who are
from day to day awaiting it (at which they might with
reason feel aggrieved), and confer it upon others, leaving
those unsatisfied to whom we are bound by promise, which
our honour obliges us to carry out. Were it not that these
motives, which we feel assured you will find just, retard
the fervour of our good will, such is the estimation in
which we hold your virtues, and the assurance we have of
your devotion towards us, that we should think *all honour
inferior to your merit.* But such being the state of things,
we would pray you to content yourself for this time with
these just excuses, awaiting a favourable opportunity to
avenge ourselves of the honour and affection which you
bear to us, and for which we shall never be found ungrateful."[22] But Elizabeth's habit of conveniently forgetting her
promises is all-too-well known.

Furthermore, the Garter regulations contain no clause
forbidding the sovereign from electing additional foreign
princes while some alien rulers still await induction. In
1593, however, Elizabeth probably would not even have
considered Frederick for membership inasmuch as he was

[22] P.R.O. SP 81/7, fol. 194. Translated in Rye, p. lxiii.

still but a minor noble. Although the situation had altered by 1594, with Frederick's succession to the dukedom, no Garter elections were held that year. Though the same five vacant stalls remained to be filled in 1595, again no elections. There were openings in the Order, then, at the time of the Breuning mission. But the Queen deliberately chose not to fill them. This action, accompanied by her denial of the promise and her hedging on the legality of electing Frederick, makes it appear as if Elizabeth in 1595 was intentionally playing the Duke—baiting him with a Garter which she would alternately dangle before and withdraw from him until she could find the propitious moment to let him snap it up. That Frederick recognized he had done nothing to warrant a favorable nod from the Queen is apparent from certain remarks Breuning made in his initial audience: "And even though His Highness has not yet accomplished aught in the service of Your Majesty wherewith to merit this grace and benefaction, he promises that he will avail himself of every opportunity to serve Your Majesty and the Illustrious Order."[23]

3) At the very time of the Breuning mission such an opportunity arose—one which the Queen cleverly refrained from mentioning until after the Feast of St. George. At the conclusion of Breuning's second audience, Elizabeth switched the conversation to certain political charges to be carried back to Frederick. Among these she stated, "Your illustrious Prince shall suffer my merchants to go and come and carry on their trade in all security, and regard all Englishmen as persons recommended to his particular protection."[24]

This charge undoubtedly stems from the strained relations which developed early in 1595 between England and the Hanse towns of north Germany over trade policies.

[23] Klarwill, p. 365.   [24] *ibid.*, p. 391.

William Camden refers to the difficulties in his Annals, noting in that year the complaint of the Hanse merchants to the German emperor "that their Priviledges about payment of Customs, granted in former times by the Kings of England, were of late infringed, their Goods taken from them in the Portugall Expedition, and Monopolies of English Merchants erected in Germany."[25] By mid-1595, feelings had become extremely bitter, and they continued to worsen over the next two years—affecting more of the German states—until Emperor Rudolph II issued an edict on July 22/August 1, 1597, barring the English Merchant Adventurers from the German Empire.[26]

It is not surprising, therefore, that Duke Frederick should suddenly assume great political value to Elizabeth in 1595. Nor is it coincidental that Lord Willoughby of Eresby apprising the Earl of Essex of the attitudes various European rulers now held toward England should write on November 26, 1595, after having just left Germany: "And here I wish, that we might embrace that prince, the

[25] *The History of the most Renowned and Victorious Princess Elizabeth* . . . [1615], 3rd ed. (London, 1675), trans. T. Hearne (1717), p. 505.

[26] There is a copy of the Emperor's proclamation which traces the Hanse-English relations from the fourteenth century, culminating in the edict, in *H.M.C. Salisbury*, VII, 307-308. The troubles first broke out in Stade, Emden, and Hamburg—the principal cities in which the English merchants had settled. The Hanse center in England was the Steelyard on lower Thames Street in London. Abundant information exists in contemporary documents to piece together accurately the Hanse-English difficulties. Indeed, these documents, some of which will be cited in the course of this study, are to be preferred to the sketchy accounts which appear in any of the standard histories, for the primary sources reveal clearly how this trade struggle permeated all levels of Elizabethan society during the years 1595 to 1604. For a sampling of the documents, see P.R.O. SP 82 (State Papers Foreign: Hamburg and Hanse Towns 1588-1659); B.M. Cott. MS. Galba D. XIII and Galba E. I; *H.M.C. Salisbury*, vols. XII, XIII *passim; Cal. S.P. Dom., Eliz., 1595-1597,* entry 28, p. 115; entry 57, pp. 122-124; entry 64, pp. 332-334; entry 171, p. 526. See also Richard Ehrenberg, *Hamburg und England im Zeitalter der Königin Elisabeth* (Jena, 1896), pp. 186-215.

king of Denmark, and the Duke of Wirtemberg, since we may tie all, as before, with a garter."[27]

Elizabeth apparently had been playing with the same notion, for we find Duke Frederick nominated to the Order in April 1596—in a year which was critical in Anglo-German relations. How accurate the observation of Peter Heylyn, who in the early seventeenth century wrote of the Order: "Yes, and sometimes the Soveraigne, as he may by statute, pronounceth him to be elected . . . whom hee conceives most worthy, *and like to be most profitable to his affaires*: as Casimire King of Poland was in the 28. of Henry the sixt, pronounced elected by the King; though he was named by one onely of the six Electors."[28] Frederick was named by one only of the twelve electors in 1596—Lord Sheffield. But that was enough; the Queen had acted through his lordship. However, Elizabeth refrained from certifying the election—proroguing it to the following year—although, as the *Blue Book* notes, everyone had expected the naming of some new Knight. Though the only record of Frederick's nomination is the 1596 Scrutiny entry in the *Blue Book*, knowing the Elizabethan courtier's

[27] Birch, *Memoirs of the Reign of Queen Elizabeth. . . ,* I, 323. The reference to the "as before" may apply to Henry IV of France, who, though elected in 1590, was not invested until 1596. Or it may refer to John Casimir, Count Palatine who became a Knight of the Garter in 1579. A letter written by Casimir to Queen Elizabeth from Heidelberg on July 27, 1591, deals with the question of the Count's favoring English merchants trading in Stade. See B.M. Cott. MS. Galba D. XIII, fols. 116-116v. This letter appears to be but one in an exchange between the two rulers. What it demonstrates is the tie which exists between political maneuvering and the award of the Garter to foreign princes.

[28] *The Historie of that most Famous Saint and Souldier . . . St. George. . . ,* 2nd ed. (London, 1633), p. 337. Italics mine. Heylyn continues: "For commonly our Kings are first well assured of the [foreign] parties good affection to them, before they choose him. . . . But other Princes, as by our Kings it is conferred upon them for an honour; so by them is it accepted also as a favour: the greatest pledge of amitie and faire correspondence between friendly Princes" (p. 338).

penchant for gossip (and in the absence of any admonition
to the contrary in the *Blue Book*), we can assume that
many an individual at Court knew that Frederick's dream
was soon to be realized. This information could easily have
filtered down to Shakespeare—either through a direct tip
from Lord Hunsdon or through the dramatist's own keen
ear for Court happenings.

Frederick apparently remained ignorant of how his
cause was proceeding, for if the absence of correspondence
since his letter of July 1595, serves as any criterion, he
seems to have held to his promise to await the Queen's
pleasure in bringing his election to fruition. But he was
far from a forgotten man in English diplomatic circles—
especially as the Hanse troubles grew more intense. Dr.
Christopher Parkins, who since 1590 had been Elizabeth's
adviser on Hanse matters, wrote to Sir Robert Cecil on February 21, 1597, to the effect that he believed direct consultation with certain German princes would best benefit the
English cause. He recommended that the Government com-
municate with these princes either through envoys or "by
letters . . . as it were by the way saluting them from her
Majesty, which may seem the fitter if there were any other
matter wherein they were now especially to be confirmed.
The fittest Princes for like occasions are Breame, Magdi-
burg, Saxon, Rhene, Hassia and Wirtenberg."[29] Two
months later the Württemberg prince was tied with a
garter.

The entire timing of the initial nomination of the Duke
with his subsequent election a year later and the rupture
in Hanse relations is much too exact to be considered
mere coincidence. And the fact that, while only one knight
nominated Frederick in 1596, all ten electors placed his

[29] *H.M.C. Salisbury*, VII, 79.

name first in the "Prince" category in the 1597 Scrutiny
only confirms the theory that Frederick was elected to the
Order out of sheer political expediency. As such, when
coupled with the image of the Duke built up among the
English over a period of years through his persistent—and
not particularly private—efforts to become a Knight of
the Garter, the final picture is one of a man who provides
a perfect target for some slight ridicule in a play whose
composition is so integrally bound up with that same
Garter.

The Mompelgard story does not stop with Frederick's
election to the Order. What remains demonstrates that the
Duke was a topical figure through 1604; that his constant
nagging for the Garter did not cease with his election,
thereby keeping the element of ridicule in the character
of the "Duke de Jarmany" intelligible to late Elizabethan
audiences; that Frederick became more entwined in the
Hanse-English quarrel; and that this quarrel would have
made the Germans so opprobrious to English audiences
that the latter would have taken a great delight in the
references to Germans as cozeners and villains in the
Merry Wives.

On April 23, 1597, Frederick became a Knight-Elect of
the Order of the Garter. On May 4, Rowland Whyte reported to Sir Robert Sidney a conversation he had had on
the subject with Lord Howard. Sidney had been fretting
over his inactivity in Flushing and had hoped that he might
get a commission to deliver the notification to Frederick.
Accordingly, Whyte broached the matter to Lord Howard.
He replied to the effect that ". . . it may be 7 Years hence,
in Comparison, er that the Queen doe send any vnto him
[the Duke of Württemberg] with it; for it is thought, that
the next Yeare some other Prince of Germany may be
chosen, and soe take our Time to send the Garter vnto

them; *but in Hast I am sure it will not be. . . .*"[30] In the
same letter Whyte reports a similar response from a Mr.
Lake about the delay in dispatching the insignia. Two
weeks later Whyte wrote again, confirming his initial report, "Vpon Monday, as I writ vnto you, the Knights goe
to Winsor to be installed, but here is no Kind of Speach of
sending the Garter to the German Prince."[31]

It took the Queen until October 10 to notify Frederick
of his election. We can well imagine his elation when he
received Elizabeth's letter of certification (P.R.O. SP 81/8,
fol. 22), for, in spite of his intentions to remain silent,
Frederick had sent a barrage of letters to England on April
30/May 10 beseeching the Queen to remember her promise
and asking Lord Burghley and Sir Robert Cecil to intercede for him.[32] A few months later he followed these up
with another letter to the Queen and a present of a beautiful chandelier (P.R.O. SP 81/8, fol. 8).

Now, neither the election of the Duke *in absentia* nor
the fact that Elizabeth waited almost six months before informing him of it can be considered irregular or in any
way counter to procedures of the Order. Article XIII of the
Statutes clearly states: "Item, it is agreed, that all strangers,
that shall be electe Felows of the seid Order, shall be certifyed by Letters of the Soverayne of their election. The
which letters of Certification, with the Statutes of the said
Order, under the common Seal shall be sente unto them at
the coste and charges of the said Soverayne, in all diligence,

---

[30] Collins, II, 49-50. Whyte paraphrases Howard's reply. Italics mine.
See also *H.M.C. Penshurst*, II, 275.

[31] Collins, II, 36. This letter of May 19 is erroneously dated by Collins
as April 9, but Kingsford has ascertained the correct date in *H.M.C.
Penshurst*, II, 282.

[32] P.R.O. SP 81/8, fols. 3, 5, 7 respectively to Cecil, Elizabeth, and
Burghley. There is also an undated letter from Breuning to the Queen
among the Salisbury papers which plays on the familiar theme of the
promise and the Duke's wish "that his hopes may be no longer deferred"
(*H.M.C. Salisbury*, XIV, 330). I conjecture the date between late 1595 and
October 1597.

and at the fardest thei shall bee certifyed of this within foure moneths after the sayd Election, to th' ende that the seid Electe may advyse them by the said Statutes, if they will receyve the sayd Order, or no. But if the said Soverayne have greate and high lettes and busynes, that then he may deferre the Certification of the said Election at his good pleasure, unto tyme of opportunitie and convenient.[33]

This Article plainly establishes the propriety of the *in absentia* election and delayed notification in receiving Frederick into the Order. Recognition of this fact should clear up the misconceptions which critics have long held about the Duke's *in absentia* election, usually interpreting it as an overt act by Elizabeth to show how little attention the English courtiers were to pay to the Duke's membership.[34]

Nor did the nobles need a delayed election notification to tell them why Frederick had been taken into the Order. In fact, there seems to be a direct correlation between dangling the Garter before the Duke and Elizabeth's need for his support among the German princes. That interval between Frederick's election and his receipt of the letter of certification, for example, was a crucial period in the Hanse relations. The edict of Rudolph II barring the Merchant Adventurers from the German Empire came on July 22/August 1, 1597. At this very time a memorandum was is-

[33] No time limit is set by this statute for investing the stranger prince with the Garter insignia once his election has been certified. Ashmole comments that ". . . it hath come to pass . . . that the Habit and Ensigns have been sent over, sometimes soon after the Stranger's Election, at other times not till some years after, and at all times when the Soveraign hath thought fit and convenient" (p. 390).

[34] See, for example, Hotson, *Shakespeare versus Shallow*, p. 114; also Samuel Tannenbaum, "Prof. Hotson's Conclusions about Shakespeare Disputed," *New York Times*, October 18, 1931, Sec. 3, p. 2. Tannenbaum attempts to refute the theory that the "Duke de Jarmany" is Duke Frederick; but from the tenor of his argument, it is clear that he wrote with no conception of Garter election procedures and has completely misinterpreted the facts in building his case.

sued by someone in English diplomatic circles headed "A memorial of Certain thinges to be done concerning the Emperors mandate against the merchant Adventurers" and directing that letters be sent by special messenger to the Emperor and various German princes, among them the Duke of Württemberg, whose name is repeated several times in the document.[35]

The Queen, in mid-1597, needed all the friendship she could get from the German princes, especially in the light of Dr. Christopher Parkins's communique of September 11 to Sir Robert Cecil: "The commissioners have met twice on Lisman's letter, but to little effect. Spain is urging divers Princes to join against England, not to suffer themselves to be hindered in their passage, and to protect their subjects from spoil. Lisman pretends to be come only in zeal to prevent further troubles. He wants the Hanse towns to have the liberty of subjects, and their former privileges. The merchants have been asked to give in their reasons to the contrary, but have not done it. . . . Yet the Queen so worded her letters to the Emperor and other Princes, that it might be conceived the Hanse towns should pay no more than her subjects; and if this be not granted, they may probably molest her. . . . There are now some Hanse towns which favor Spain and are hostile to Her Majesty. . . ."[36]

Within the month the letter of certification was sent to Frederick, with a follow-up note dispatched on November 10.[37] Two days after the first letter went, on October 12,

---

[35] P.R.O. SP 82/4, fols. 73-74. The MS is conjecturally dated by the P.R.O. as July 22, 1597.

[36] Cal. S.P. Dom., Eliz., 1595-1597, pp. 499-500.

[37] P.R.O. SP 81/8, fol. 24. It is interesting to note that a similar conjunction of political maneuvering and a Garter ceremonial surrounds the investiture of Henry IV of France in the autumn of 1596. Henry had been elected to the Order in 1590. Ashmole (p. 383) notes that Elizabeth gave him speedy notification of his election. But the Queen took no further steps to send the insignia to Henry until 1596. The time she did pick was one during which she was having difficulties with Spain and therefore was engaged in negotiating a mutual assistance treaty with France. The

Lord Burghley received word from one Ludolph Engelstedt that "by a proclamation of the Emperor of Germany, all English merchants are to depart the empire within three months, on pain of confiscation and imprisonment, so that the Merchant adventurers will have to leave Stade . . . . The merchants are much troubled with this proclamation and want directions. . . ."[38] The situation had deteriorated so much that Stephen Le Sieur, one of Eliza-

---

manner in which the treaty and investiture are constantly referred to in contemporary documents leaves little doubt that there was deliberate planning behind the entire business. The Signet Office Docquet Books, for example, bear two contiguous entries for September 3, 1596: "A letter in french to the king concerning his oath for obseruacion of the league. . ." and "A like letter to the said king by the Earle of Shrewsbury for accepting the Garter. . ." (Index 6800, fol. 607v). And an account of the delivery of the Garter to Henry in B.M. Harl. MS. 1355, fol. 24 reads: "On Satturday the 9th the oathe of Confederation betwixt the Kinge & the Queenes Majesty of England was very Solemplye taken in the Church of St Owyn in the presence of the french Nobillitie Lord spirittuall & temporall who that day kept the right hand of the Quire.

"On Sunday next followinge beinge the xth of October the order of the Garter was most Royally performed in the said church. . . ." See further, John Stow, *Annales*. . . (London, 1631 ed.), pp. 777-782.

[38] *Cal. S.P. Dom., Eliz., 1595-1597*, p. 515. An indication how widespread this ban was and how it was received by the German rulers may be gathered by sampling some of the entries in Thomas Rymer's *Foedera*: October 8, 1597. "The magistrates and council of Embden assure the Q. that they would wish the merchant adventurers to trade at Embden, but that town is included in the emperor's edict ordering them to leave the empire." (II, 826) December 1, 1597. "John Adolphus, duke of Schleswig, to the Q. He cannot avoid obeying the emperor's edict, that the English merchants must leave the empire." (II, 826) December 21, 1597. "Otto duke of Brunswick to the Q. expressing his regret at the above-mentioned edict." (II, 826) Some of the English attempts to woo other princes may be gathered from a letter of the Lord Chamberlain, Lord Hunsdon, to the Earl of Essex, dated November [9?] 1597: "Her Majesty was pleased with your good assurance of the Landgrave's profession to continue her's, and to relieve her subjects, notwithstanding the Emperor's late mandate. I let her know your advice that she should send to the Duke of Wirtemberg, and the Palsgrave of the Rhine, to join in that course which you thought would prevail, and occasion them to account the mandate as not proceeding from the empire, but from the House of Austria. Her Majesty seemed to have sent to those Princes already to the like effect, yet notwithstanding, by the gentleman that is now to return from you, she will return thanks for the honourable offer of the Landgrave." (*Cal. S.P. Dom., Eliz., 1595-1597*, p. 529.)

beth's ambassadors, wrote to Sir Robert Cecil on November 16, 1597: "I hear that the Imperial Diet is to begin the 20th of next month. The mandate against the Merchant Adventurers is proclaimed in the dominions of the Landgrave of Hesse, and of most of the Princes of the Empire. In Lubec, seizure is already made upon our English merchants' goods, and the like measure is doubted in Stade, for some Dutchmen in those parts begin to deny payments to English merchants. If this vehement course takes place against them, it will ruin many of Her Majesty's loyal subjects, *resident as well here in England* as beyond seas."[39]

How embittered the English must have become toward the Germans in the late 1590's. If Mompelgard's secretary Jacob Rathgeb found the Londoner of 1592 sufficiently hostile to cause him to comment in his *Badenfahrt*, ". . . they care little for foreigners, but scoff and laugh at them; and moreover one dare not oppose them, else the street boys and apprentices collect together in immense crowds and strike to the right and left unmercifully without regard to person. . . ,"[40] we can imagine the state of mind of that same Londoner—whether he be commoner or courtier—when the commercial prosperity of his nation was threatened by the upstart Hanse. These Hanse had even gone so far in 1597 as to try to get Poland and Danzig to join with them in stopping all trade with England; to forestall such a move Elizabeth sent Sir George Carew on an embassage to these countries.[41] The Londoners, then, must have gleefully received the Queen's edict of January 13, 1598, ejecting all Hanse merchants from England.[42]

---

[39] *Cal. S.P. Dom., Eliz., 1595-1597*, p. 534. Italics mine.

[40] Rye, p. 7.

[41] Camden, *The History of the most Renowned and Victorious Princess Elizabeth*, pp. 537-539.

[42] Addressed to the Lord Mayor and Sheriffs of London, the Queen's proclamation serves as another illustration of how the Hanse troubles became a vital part of the daily life of all levels of English, and particularly London, society: "A mandate has been issued from the Emperor, to

Relations do not seem to have improved during the first five months of 1598; in fact, the negotiations which had commenced the previous December apparently came to naught, [43] for on May 17, 1598, the Earl of Essex, in writing to one of the lords, remarked, "The Emperor hath exiled our nation from all trade within the empire. The Hanse

---

all electors, prelates, and other officers and subjects of the Empire, reciting sundry complaints made by the Hanse Towns of injuries committed against them in our realm, and complaints against our Merchant Adventurers, without hearing of any answer made to the said Hanse Towns in disproof of their complaints (the same being most notoriously unjust and not to be maintained by any truth). Yet nevertheless, by that mandate, our English merchants, namely, the company of Merchant Adventurers, are forbidden any traffic within the empire, and commanded to depart upon great pains, and to forbear all havens or landing places, or any commerce by water or land in the empire, on pain of apprehension and confiscation. We have sent letters to the Emperor, and electors, etc., requiring to have the same mandate revoked or suspended, yet being uncertain what shall follow thereupon, we have thought it agreeable for our honour, meantime to command that all in our realm appertaining to the said Hanse Towns, and especially such as reside in London, either in the house called Stillyard, or any other place, do forbear to use any manner of traffic, or make any contracts, and do depart out of our dominions, in like sort as our subjects are commanded to depart out of the empire, and upon the like pains.

"For the execution of this our determination, we will that you, the mayor and sheriffs, forthwith repair to the Stillyard, and give those who reside there knowledge of this our command, charging them that by the 28th of this month, being the day when our merchants are to avoid from Stade, they depart out of this realm; you are also to give knowledge hereof to those of the Hanse Towns belonging to the empire, remaining in any part of our realm, to depart by the same day; and you, the Mayor and sheriffs, calling the officers of the Customs to you, are to take possession of the said house on the 28 instant, to remain in our custody, until we shall understand of any more favourable course taken by the Emperor for restitution of our subjects to their former lawful trade within the empire" (*Cal. S.P. Dom., Eliz., 1598-1601*, pp. 5-6).

In informing Sir Robert Sidney of the serving of this order on January 14, Rowland Whyte noted, "They [the Hanse] stoode much vpon the Priuiledges of the Stilliard; but they see yt serue to smale Purpose" (Collins, II, 81). Still, Elizabeth did grant a brief stay in the execution of the edict.

[43] See *Cal. S.P. Dom., Eliz., 1595-1597*, entry 37, p. 543; entry 49, p. 548; entry 65, p. 553; entry 71, p. 554; also *Cal. S.P. Dom., Eliz., 1598-1601*, entry 17, p. 6; entry 26, p. 11; entry 29, p. 13; entry 9, pp. 49-50.

towns are our professed enemies."[44] And on that note matters remained more or less stalemated until 1602. Elizabeth's attention was taken up with other problems, notably
the Irish rebellion of 1599.

Oddly enough, about the very time Essex was summing
up the Anglo-Hanse position, Duke Frederick and the
Garter again clamor for attention. Though Frederick had
been elected to the Order, he had not received any of the
ensigns—that is, he was still to be invested as a Knight-
Elect. Accordingly, he sent Benjamin von Buwinckhausen,
the able second to Breuning in the 1595 mission, to London to thank Elizabeth for his election and to try to obtain
the desired insignia.[45] But in vain. So Frederick resumed
his badgering by letter. He wrote on November 10/20,
1598, and again on January 10/20, 1599, begging the
Queen to send him the ensigns.[46] Knowing Elizabeth's
reputation for frugality, we may wonder whether the sheer
cost of fashioning these ensigns and dispatching an investing delegation dissuaded her from taking any prompt
action.[47] At any rate, the Queen's delaying tactics served to

[44] *H.M.C. Salisbury*, VIII, 170. The name of the addressee has not been
preserved.

[45] There are two letters on the subject which Buwinckhausen wrote in
May, 1598, while he was in London, to Sir Robert Cecil. See B.M. Cott.
MS. Galba D. XIII, fols. 179-180. Frederick also had sent a delegate to
the Queen a couple of months earlier for some reason as a warrant of
payment, dated March 24, 1598, to a goldsmith named Hugh Kayle in-
dicates. Kayle fashioned a gold chain which he delivered to "Adam
Viman a gentleman sent from the Duke of Wetemberge." See P.R.O.
E404/133.

[46] P.R.O. SP 81/8, fols. 85, 94.

[47] It cost Elizabeth £440 alone "for one George of gould garnished with
Diamondes, and one Garter of purple velvett imbrodered with letters of
gould, and garnished with buckelle and pendante of gould, sett with
Rubies and Diamondes. . .to be sent to our Deere brother the French
kinge" (Warrant of Issue, P.R.O. E404/132). About another thousand
pounds would be a fair estimate for the cost of transporting, feeding, and
housing the investiture delegation. Having just sent an embassage to
France in 1596, the Queen may have been in no hurry to duplicate such
expenditures for Frederick. Installation fees for foreign knights also were
payable by the sovereign (Ashmole, p. 463).

keep the Duke a useful servant in the Hanse affair, for he is found writing on February 25/March 7 that he promises to mediate with the Emperor and princes over the banning of the Merchant Adventurers.[48] Still, Frederick temporarily became balky, as Stephen Lesieur, Elizabeth's ambassador to the Assembly of German Protestant Princes, found. Lesieur, writing to Sir Robert Cecil from Spire on April 29/May 8, 1599, reported: "The duk of Wirtemberg, his manner of intertaining me, & speech in favour of the Spanishe proceedings in th Empire, hath ben strange and contrary to my expectacion, the one I impute, for that he hath not the ordre of the garter which he greatly desireth, & wherof with his owne hand he writtes himselffe knight, the other for that he is in treatie with the Emperor."[49] This is the last word about Duke Frederick and the Order of the Garter that we have during Elizabeth's reign.[50]

And it parallels the *status quo* in the Anglo-Hanse negotiations. These show no signs of resumption until early 1602.[51] Their period of quiescence—1598 to 1601—means that any possible anti-German allusions in drama would have been eagerly received by the stranger-hating Londoners throughout the period. Though Shakespeare may have designed his slight ridicule of the "Duke de Jarmany" especially for the performance at the Garter Feast of 1597, the course of history gave his German references new relevancy over the next few years, making them particularly meaningful to a popular audience. What a reaction from

[48] P.R.O. SP 81/8, fols. 113-116v.

[49] P.R.O. SP 81/8, fol. 133.

[50] Other correspondence is extant, but contains no mention of the Order of the Garter. Frederick apparently resigned himself to waiting in silence. Buwinckhausen, writing to Sir Robert Cecil from Paris on April 8/18, 1602, concludes his letter with a plea to ". . . keep in her [Elizabeth's] remembrance a Prince who is so well-disposed to her Government" (*H.M.C. Salisbury*, XII, 100).

[51] See the many references in *H.M.C. Salisbury*, Vol. XII.

the groundlings to those lines of Bardolph's in which he replies to the Host's question, "Where be my horses?" with

Run away with the cozoners. for so soone as
I came beyond *Eaton*, they threw me off, from behinde
one of them, in a slough of myre; and set spurres, and
away; like three *Germane*-diuels; three *Doctor* Faustasses.

(IV.v.67-71)

And Buwinckhausen's presence in London during the spring of 1598 would have made the "Duke de Jarmany" references fresh once again, implanting them more indelibly in the memories of the theater-goers. Of course, in the absence of records of performance one can do no more than establish the international climate against which any production of the *Merry Wives* would have been played. But, as previously shown, there is reason to believe that the play was in the repertory during this period.

How topical would the references have been when friendly relations were reestablished with the Germans? Amenity, once serious negotiation recommenced in the fall of 1602, was not fully achieved for at least another year, though the death of Elizabeth in March 1603, almost deadlocked the discussions which had been under way in Bremen.[52] But matters seem to have worked themselves out by the following October. At this time the Emperor Rudolph is found writing to James I from Prague declaring friendly intentions toward the new ruler (B.M. Cott. MS. Galba E. I, fols. 119-119v). He also issues a declaration for putting an end to the Hanse dispute (B.M. Cott. MS. Galba E. I, fols. 121-122). With the resumption of trading

[52] See entries in Thomas Rymer, *Foedera . . . of the Documents Relating to England . . .* (London, 1869-85) for August 30 and September 26, 1602 (II, 829) and for April 16, 1603 (II, 831). Also *H.M.C. Salisbury*, xv, 13-15, which is a letter of March 27, 1603, from the English commissioners in Bremen to the Privy Council briefing that body on the progress of the negotiations.

privileges for the Merchant Adventurers, a Jacobean thea-
ter-goer may have reacted less sharply to any German allusions he heard on the stage, but it is doubtful whether he
would have entirely eradicated any possible overtones of
the Hanse affair from his memory.

And if the *Merry Wives* were revived by the King's
Men any time between July 1603, and late 1604, the "Duke
de Jarmany" allusions would have taken on a fresh life,
for Duke Frederick resumed his campaign to receive the
Garter ensigns soon after James came to the throne. The
tactics he adopted—though undoubtedly proper in broach-
ing the matter to a new sovereign—were the same that first
brought him notoriety in Elizabeth's court: letters and an
embassage. Writing to James at the beginning of July (B.
M. Harl. MS. 1760, fols. 90-90v), Frederick extended best
wishes to him on assuming the crown and indicated a hope
that James would favor him with the honors Elizabeth had
promised. The Duke then dispatched Buwinckhausen to
England to further his request for investiture. The envoy
arrived on July 13, was granted an audience with the King
on the 30th, and left for Württemberg on August 7.[53]

Buwinckhausen's mission proved successful, for on September 18 James signed and sealed the long-sought letter
of authorization for the investiture of Frederick, Duke of
Württemberg and Teck, Count Mompelgard with the
Garter robes and ensigns.[54] To this end, he appointed
Robert Lord Spencer of Wormleiton to head the investing
delegation and notified Frederick of his action in a letter
dated September 24 (B.M. Cott. MS. Galba E. 1, fol. 11).
The investiture ceremonies took place in Stuttgart on November 6, 1603.[55] Frederick rewarded the English en-

[53] *H.M.C. Salisbury*, xv, 182; Bodl. Ashm. MS. 1115, fols. 76-76v.
[54] A copy of the letter has been included in Cellius, pp. 133-134. See
also pp. 115 and 117 for Cellius' account of the success of the Buwinckhausen mission.
[55] See accounts in Stow, p. 828 and in Cellius, pp. 129-163, 229-260;

tourage handsomely.[56] In accordance with Garter procedure, he also sent letters to James indicating that he would dispatch a proxy delegate for his installation the following St. George's Day.[57]

The envoy chosen as Frederick's proxy was Count Philipp von Eberstein. He arrived in London on April 15, and gained an audience with James three days later. The King told him, "I shall do whatever I can in honor of the Duke your master. To-morrow you must go to Windsor to the Installation. Three Knights of the most noble Order of the Garter shall accompany you, and on Monday next we shall celebrate the Feast of Saint George in this our city of London."[58] As the King commanded, von Eberstein left for Windsor the following day, the 19th. And on April 20 in St. George's Chapel—where his master, Duke Frederick, twelve years earlier had stood and longingly glanced at the shields, helmets, and banners adorning the stalls of the Knights of the Garter—von Eberstein took the oath of the Order.[59] A few days later at the Grand Dinner on St.

---

portions of this latter have been translated in Ashmole, pp. 411-416. Lord Spencer's records of the entire mission have been preserved in his family archives and are described in the Appendix of the *Second Report of the Royal Commission on Historical Manuscripts* (London, 1871), p. 20. These records indicate that the total costs for the journey were £948-16-2.

[56] Cellius, p. 259, enumerates the gifts.

[57] This action was taken in accordance with that section of Article XIII of the *Statutes* which states: "After that the Certification have byn delivered, and that the Soverayn shall be certefied, that the said Electe will receyve the said Order; Then the Soverayne shall sende unto the said Electe, by his Ambassadours, his hole habit, with the Garter and Coller. And that all suche straungers, of what estate, dignytie, or condicion that thei be of, shall sende within vii. monethes after the reception of the said Gartier, Coller and Habit, and that he have certified the Soverayn to have receyved those thyngs, a sufficient Deputie or Attorney after th' estate of his Lorde and Maister, so be that he be a Knyght without Reproche, to be stalled in his place. . . ."

[58] Quoted by Rye, p. lxxxiv, from Sattler, *Geschichte des Herzogthums Würtenberg unter der Regierung der Herzogen* (1772), Theil V. Rye, pp. lxxxiii-lxxxiv, reprints Sattler's account of von Eberstein's mission.

[59] A contemporary account of the proxy installation has been preserved

George's Day, 1604, at Whitehall, von Eberstein heard the stiles of Frederick proclaimed.[60] Thus the Duke's long and steadfast pursuit of the coveted Garter came to an end.

By a strange coincidence, the Garter ceremonial at which Frederick finally became a full-fledged Knight-Companion of the Order was also the one—upon completion of the proxy installation service—at which the death of George Carey, Lord Hunsdon was commemorated. In such manner the essences of the two Garter knights whose names have been so closely linked with the composition of the *Merry Wives* met for an instant in the most appropriate of places, St. George's Chapel, Windsor.

Several months after the installation of Frederick, James had the opportunity of seeing a performance of the play, for the *Merry Wives* received a Court production on November 4, 1604. Whether he smiled at the sly allusions to the "Duke de Jarmany" we can only ponder. As for Duke Frederick, little is heard of him in England between 1604 and his death four years later.

---

in Bodl. Ashm. MS. 1108, fol. 81 and is herewith cited: "The 19. of Aprill 1604. The Graue Van Evestan accompanied with the Earles of Worcester, penbrock and the Lord Scroope ioynt Commissioners to enstall him for Frederick Duke of Wirtenberg came to Windesor and were lodged in the Deane and prebendaryes houses.

"The day following after 9. of the clock the Almes Knightes and Heraldes brought the Commissioners from their Lodging apparelled in their Robes to the Deanes house where hee lodged.

"From thence they proceeded to the Chapter-house placeing the Graue in a Chaire behinde the high Aulter.

"From whence they came to the Quier and first standing before their Stalls, afterwardes ascended. When the Anthem was ended, they returned to the Chapter house and Garter bringing from thence the Roabes and Coller vpon a Cushion, they tooke the Graue with them to the Quier, where the deane ministred the Oathe to him standing in the midle Seat. Then was he led vp to his Seate, and the Roabes layd vpon his arme.

"After the first service the Llordes the younger first came downe from their places, and then the Banner of ye Lord Hunsdon was delivered to the Erles of Penbrook & Worcester who offered up ye same & accordingly the Sword & Helmett ye Officers preceding."

60 Von Eberstein's presence is listed in the *Blue Book*, p. 154.

In linking the historical Duke with his fictional counter-part, we note that critics have long detected a play on the name Cousin Mompelgard in that odd Quarto phrase *cosen garmombles*. Rye (p. xcviii) saw *garmombles* as "a play-ful joke upon the Duke's title. . . ." Hotson (*Shakespeare versus Shallow*, p. 115), on the other hand, believes it a scrambling of Garter and Mompelgard.[61] The word *cosen* is assuredly a pun, for Elizabeth constantly addresses the Duke in her letters as "Mon Cousin," a salutation which James I also used. And the very towns which these Ger-mans cosened were visited by Count Mompelgard during his 1592 journey (Rye, p. 11), though this fact may merely be coincidence. The last point about a German duke who does not come to court for the celebration fits, of course, exactly with the Garter election of Frederick *in absentia*—an entirely proper procedure for foreign rulers and one that since it was not secret, Shakespeare could have been expected to be acquainted with.

While the "Duke de Jarmany"–"cosen garmombles" sa-tire may have been intended solely for the original Court performance of the *Merry Wives*, political events caused the entire incident in which it is embedded—the horse-steal-ing subplot—to become especially meaningful to the pop-ular audiences who might see the play either in London or in the provinces. This came about through the rupture in Anglo-Hanse trade relations. The fact cannot be overlooked that there is a direct parallel between the ebb and flow of negotiations between Emperor Rudolph and Queen Eliza-beth and the action taken in awarding Duke Frederick his

---

[61] Attempts have been made to link the word with the dialect term *girmumble*, a mess, but no convincing linguistic evidence has been offered. See J. Douglas Bruce, "Two Notes on 'The Merry Wives of Windsor,'" MLR VII (April 1912), 240-241; Crofts, pp. 41, 161, n.46; Greg, *The Shake-speare First Folio*, p. 337, n.4. Rye, we must remember, in theorizing *garmombles* as a metathesis of syllables in Mompelgard, was writing long before Hotson established a link between the Duke and the Garter.

Garter. And it must not be forgotten that Frederick was elected to the Order in April 1597, just three months before these negotiations reached their nadir with the issuance of Rudolph's edict of expulsion for the Merchant Adventurers. How much of this political maneuvering may have trickled down to the groundlings one cannot say. But what is evident is that the popular audiences would have pounced upon the *Merry Wives* portrait of Germans as horse thieves whereas "those in the know" would have had the added enjoyment of recognizing the dig at the persistent Duke Frederick. After all, from the number of people written to and seen in his behalf in both high and low circles during his quest for the Garter, there can be no doubt that he was a sufficiently recognizable figure to have made any topical references to him intelligible for a good many years. In fact, the number of cognoscenti would have been significantly increased as a result of the physical intermingling between the Londoners and members of the successive missions the Duke sent to England in 1595, 1598, 1603, and 1604.[62] Even the succession of James to the throne with the subsequent settling of the Hanse affair as well as the investiture and installation of Frederick in the Order would not have materially affected the topicality of the German references—though after 1604 the story behind the allusions would have gradually faded until mere dialogue remained.

[62] Cellius (pp. 107-109) reports that Buwinckhausen, assisted by Christopher ab Haugwitz, came to England in 1600 to request the ensigns of investiture. I have found no mention of this mission in any other source. Since Cellius does not note the 1598 mission of Buwinckhausen, it is possible that he has misdated it as occurring in 1600.

CHAPTER VIII ✦ THE HORSE-STEALING SUBPLOT

ONE of the puzzling features of the *Merry Wives* is the fourth act horse-stealing subplot. Consisting of the very brief third scene and thirty lines in the fifth, it comes out of nothing and goes nowhere. As a perusal of the parallel Quarto and Folio texts below reveals, the corrupt Quarto, aside from bringing to light the reading "cosen garmombles," merely shows garbling of the Folio version and makes no contribution toward integrating this episode within the play.

Parallel Text Horse-stealing Scenes

Q 1232-42	F IV.iii.1-12
Enter Host and Bardolfe.	*Enter Host and Bardolfe.*
Bar. Syr heere be three Gentlemen come from the Duke the Stanger [*sic*] sir, would haue your horse.	*Bar.* Sir, the Germane desires to haue three of your horses: the Duke himselfe will be to morrow at Court, and they are going to meet him.
Host. The Duke, what Duke? let me speake with the Gentlemen, do they speake English?	*Host.* What Duke should that be comes so secretly? I heare not of him in the Court: let mee speake with the Gentlemen, they speake English?
Bar. Ile call them to you sir.	*Bar.* I sir? Ile call him to you.
Host. No *Bardolfe*, let them alone, Ile sauce them: They haue had my house a weeke at command, I haue turned away my other guesse, They shall haue my horses *Bardolfe*, They must come off, Ile sawce them. *Exit omnes.*	*Host.* They shall haue my horses, but Ile make them pay: Ile sauce them, they haue had my houses a week at commaund: I haue turn'd away my other guests, they must come off, Ile sawce them, come. *Exeunt*

Q 1344-71	F IV.v.64-94
Enter Bardolfe.	
Bar. O Lord sir cousonage, plaine cousonage.	*Bar.* Out alas (Sir) cozonage: meere cozonage.

Host. Why man, where be my horses? where be the Germanes?

Bar. Rid away with your horses: After I came beyond Maidenhead, They flung me in a slow of myre, & away they ran.

Enter Doctor.

Doc. Where be my Host de gartyre?

Host. O here sir in perplexitie.

Doc. I cannot tell vad be dad, But begar I will tell you van ting, Dear be a Garmaine Duke come to de Court, Has cosened all de host of *Branford*, And *Redding*: begar I tell you for good will, Ha, ha, mine Host, am I euen met you?

Exit.

Enter Sir Hugh.

Sir Hu. Where is mine Host of the gartyr? Now my Host, I would desire you looke you now, To haue a care of your entertainments, For there is three sorts of cosen garmombles, Is cosen all the Host of Maidenhead & Readings, Now you are an honest man, and a scuruy beggerly lowsie knaue beside: And can point wrong places, I tell you for good will, grate why mine Host.

Exit.

Host. I am cosened *Hugh*, and coy *Bardolfe*, Sweet knight assist me, I am cosened.

Exit.

Host. Where be my horses? speake well of them varletto.

Bar. Run away with the cozoners. for so soone as I came beyond *Eaton*, they threw me off, from behinde one of them, in a slough of myre; and set spurres, and away; like three *Germane*-diuels; three *Doctor Faustasses.*

Host. They are gone but to meete the Duke (villaine) doe not say they be fled: *Germanes* are honest men.

Euan. Where is mine *Host*?

Host. What is the matter Sir?

Euan. Haue a care of your entertainments: there is a friend of mine come to Towne, tels mee there is three Cozen-Iermans, that has cozend all the *Hosts* of *Readins*, of *Maidenhead*; of *Cole-brooke*, of horses and money: I tell you for good will (looke you) you are wise, and full of gibes, and vlouting-stocks: and 'tis not conuenient you should be cozoned. Fare you well.

Cai. Ver' is mine *Host de Iarterre*?

Host. Here (Master *Doctor*) in perplexitie, and doubtfull de-lemma.

Cai. I cannot tell vat is dat: but it is tell-a-me, dat you make grand preparation for a Duke *de Iamanie*: by my trot: der is no Duke that the Court is know, to come: I tell you for good will: adieu.

Host. Huy and cry, (villaine) goe: assist me Knight, I am vndone: fly, run:[&c. &c.]

[152]

Though intelligible in itself, the fragmentary nature of this subplot in terms of the over-all play is such that it has teased critics to seek for clues elsewhere in the comedy to tie up the loose threads. Dangling temptingly before them have been the two revenge schemes which are launched in the play: that of Pistol and Nym against Falstaff when he asks them to carry the love letters to Mistress Page and Mistress Ford (I.iii.) and that of Dr. Caius and Evans against the Host for sending them to opposite parts of the town for their duel (III.i.). It is upon these schemes that critics have sought to embroider the horse-stealing episode. Their job has been complicated slightly by the several scribal and/or compositorial errors in the pertinent passages, particularly in the lack of consistency between singular and plural forms in IV.iii.[1]

Noting in his introduction to the Griggs Quarto facsimile (p. ix) "that something is wanting to render this part of the play intelligible," P. A. Daniel cautiously speculated that there may have been some link (now "irrecoverably lost") between the Caius-Evans plan and the horse-stealing incident. He even went so far as to suggest that Pistol and Nym may have impersonated two of the cosen-Germans. (He had no candidate for the third.) These ideas, casually sown by Daniel, have taken root and grown into what may generally be termed a lost-scene theory. Proponents of this view believe that the *Merry Wives* originally contained a scene or passages which revealed the

---

[1] In tracing the transmission of text for these scenes, Brock (II, 383) comments, "This entire episode of the Germans and the horses is confused, of course; and I see no reason to make the nouns and pronouns consistently plural. Bardolph may quite well have been speaking of the spokesman for the group, whereas the Host was thinking in terms of the group. Yet all subsequent eighteenth-century editors and all modern editors agree in following Capell's emendation [of *"German desires"* to *"Germans desire"* and the F3 editor's emendation of *him* to *them* in IV.iii.9]." See also Brock's Appendix, List 6, Items 103-105 for textual emendations in the horse-stealing subplot.

actual planning for the trick played on the Host; however, through some form of textual revision the pertinent material simply disappeared from the script. H. C. Hart, for example, expanded Daniel's suggestion into a full-blown reconstruction of what he considered was the original episode.[2] According to his theory of text, it was lost when the F version was revised. Hart postulated that Caius and Evans, in order to execute their plan for revenge upon the Host, got Pistol, Nym, and Rugby to serve as the three coseners. Nym and Pistol participate since they are desirous of getting even with the Host for his encouraging Falstaff to dismiss them; Rugby joins in under orders from Caius because he was a witness to the prank played on his master.[3] The New Cambridge editors (p. xiv), on the other hand, dismiss the involvement of Nym and Pistol and posit that in the lost scene Caius and Evans, aided by Bardolph, carry out this revenge plot against the Host. Why Bardolph should have become an accomplice they offer no suggestion.

These views illustrate what a knotty problem it is to reconcile the horse-stealing incident with the main body of the *Merry Wives*. Actually, these attempts to interweave either or both of the revenge plots with the horse-stealing episode by positing a lost linking scene are based on noth-

[2] Arden edition, pp. lxxii-lxxvii.

[3] A somewhat similar reconstruction was made for the production of the *Merry Wives* which I witnessed at the Stratford, Ontario, festival, in August 1956. Disturbed by the fragmentary nature of the horse-stealing plot, Michael Langham, the director, adopted the interpretation that there once had been an actual scene in which Caius and Evans plotted their revenge on the Host. To carry out their plan, the two convinced Pistol, Nym and Rugby to join them. With this "restoration" in mind, Mr. Langham commissioned the Canadian playwright Robertson Davies to write additional dialogue for the script. Accordingly, the three pseudo-Germans are presented to the Host as Hogen Mogen von Kammerpot, Hogen Mogen von Arsfusstritt and Hogen Mogen von Rumplick. They mutter to the Host such choice Germanic utterances as "Gesundheit." For additional commentary, see Herbert Whittaker, "Full Shakespeare Texts Return—with Bonuses," *Toronto Globe and Mail*, August 11, 1956, p. 22.

ing more than the good intentions of the editors, for the text itself offers no justification for making such amalgamations. Examination of each of the revenge schemes will illustrate this.

The Pistol-Nym plan gets under way in I.iii, is carried into operation in II.i, and fades from the play in II.ii. At the opening of I.iii, Falstaff reveals that he is short of money:

Fal. Mine *Host* of the *Garter*!

Ho. What saies my Bully Rooke? speake schollerly, and wisely.

Fal. Truely mine *Host*; I must turne away some of my followers.

Ho. Discard, (bully *Hercules*) casheere; let them wag; trot, trot.

Fal. I sit at ten pounds a weeke.

Ho. Thou'rt an Emperor (*Cesar, Keiser* and *Pheazar*) I will entertaine *Bardolfe*. . . .

Falstaff, it will be noted, states that he must turn away *some*, not all of his retainers and appears satisfied with the Host's solution to reduce his entourage by one. After the Host and Bardolph have left the room, Falstaff reveals to Pistol and Nym that he has conceived the plan of wooing Mistress Page and Mistress Ford so as to obtain money from them:

> . . . I will be Cheaters to them both, and
> they shall be Exchequers to mee: they shall
> be my East and West Indies, and I will trade to
> them both: Goe, beare thou this Letter to Mistris
> *Page*; and thou this to Mistris *Ford*: we will thriue
> (Lads) we will thriue.
>
> (I.iii. 77-82)

From this last line it is obvious that Falstaff is still thinking in plural terms and that he has not the slightest intention

of turning away Pistol and Nym. For some strange reason
these two cony-catching rogues suddenly rebel at playing
intermediaries for Falstaff. He in turn flies into a rage at
their refusal to bear the letters, and thunders:

> Rogues, hence, auaunt, vanish like haile-stones; goe,
> Trudge; plod away ith' hoofe: seeke shelter, packe:
> *Falstaffe* will learne the honor of the age,
> French-thrift, you Rogues, my selfe, and skirted *Page*.

So Falstaff storms out accompanied by his page Robin,
abandoning the two ungrateful parasites. Unable to accept
the banishment which they have brought upon themselves,
they decide to seek revenge on Falstaff, not with "wit or
steel" but by the safer and baser course of denouncing him
to Page and Ford. This they do in II.i. At this point Nym
disappears from the script (except for an oblique reference
in IV.v.31-39). The following scene opens with Pistol attempting to patch up the quarrel. In recapitulating their
relationship, Falstaff confirms that the only reason he has
turned away Pistol and Nym is their refusal to bear the
letters—a mark of ingratitude after all he had done for
them. He reiterates that they "hang no more about mee."
Pistol accepts the verdict calmly.

Here Mistress Quickly interrupts to bring word of the
first assignation. She desires to speak with Falstaff alone,
but he tells her, "I warrant thee, no-bodie heares: mine
owne/people, mine own people." It is doubtful if he would
utter this line if he were not reconciled with Pistol. The
latter, intrigued by what he sees and hears of Mistress
Quickly, leaves in pursuit of her. This is the last we see of
Pistol. Absolutely no bitterness or hint of further revenge
tinges his departure.

From the foregoing analysis it is clear that the Pistol-
Nym revenge scheme is a coherent whole in the script.

Pistol and Nym are responsible for their own dismissal; they take their revenge in the petty manner consistent with their characters; and when efforts at reconciliation with Falstaff fail, they resign themselves to the situation. Falstaff has finished with them; Shakespeare has finished with them (dramatically, they set the countermovement of the plot into operation); they disappear. Thus all attempts to readmit them in the horse-stealing episode must be considered as reading-in to the play.

The Caius-Evans plot presents a more difficult case since it contains a direct threat by the two men to seek revenge on the Host. The quarrel between Caius and Evans starts, it will be remembered, through a challenge the doctor sends to Evans when he discovers that the latter has asked Mistress Quickly to intercede with Anne Page to accept Slender as a suitor. The Host is to make arrangements for the duel. What more ill-conceived match could there be than this between two old friends—the peace-loving parson and the quick-tempered doctor? The Host realizes this and looks upon it as sport. Page also realizes the ridiculousness of this match, for when he is invited to join the onlookers, he states, "I had rather heare them scold, then fight."

This peace motif pervades the balance of the duel incident and stands behind the Host's trick in keeping the two men separated. When the onlookers go to the field where Caius has been waiting, Shallow answers the impatient doctor's statement that Evans has not shown up with the comment,

> He is the wiser man (M. Docto[r]) he is a curer of
> soules, and you a curer of bodies: if you should
> fight, you goe against the haire of your professions:
> is it not true, Master *Page*?
>
> (II.iii.40-42)

A few lines on, Shallow adds:

> ... M. Doctor *Caius,*

I am come to fetch you home: I am sworn of the peace:
you haue show'd your selfe a wise Physician, and Sir
Hugh hath showne himselfe a wise and patient Church-
man: you must goe with me, M. Doctor.

(II.iii.54-58)

Though the Host does bait Caius a bit in this scene, it is
more in the spirit of having him work the choler out of his
system. Finally the Host convinces the doctor to sheathe his
sword, and, with the promise that he will take him to see
Anne Page, he gets Caius to leave with him. In the meantime, Shallow, Slender, and Page go off to observe Evans
who has been trying to bolster up his courage on the other
side of town. With the entry of the Host and doctor, the
peace motif reestablishes itself as the two adversaries face
each other. The Host immediately commands, "Disarme
them, and let them question: let them keepe their limbs
whole, and hack our English" (III.i.79-80). At this point
the Host reveals his trick, stating:

> Peace, I say: heare mine Host of the Garter,
> Am I politicke? Am I subtle? Am I a Machiuell?
> Shall I loose my Doctor? No, hee giues me the
> Potions and the Motions. Shall I loose my Parson?
> my Priest? my Sir *Hugh?* No, he giues me the
> Prouerbes, and the No-verbes. Giue me thy hand
> (Celestiall) so: Boyes of Art, I haue deceiu'd you
> both: I haue directed you to wrong places: your
> hearts are mighty, your skinnes are whole, and
> let burn'd Sacke be the issue: Come, lay their
> swords to pawne: Follow me, Lad of peace, follow,
> follow, follow.

(III.i.102-113)

Thus it appears that the Host's plan was derived not out of mockery, but out of a desire to heal the breach between the two. But Caius and Evans are somewhat ruffled by the trick, and to save face, decide that they must take revenge upon the Host:

> *Cai.* Ha'do I perceiue dat? Haue you make-a-de-sot
>   of vs, ha, ha?
>
> *Eua.* This is well, he has made vs his vlowting-stog:
>   I desire you that we may be friends: and let vs
>   knog our praines together to be reuenge on this
>   same scall scuruy-cogging-companion the Host of
>   the Garter.
>
> *Cai.* By gar, with all my heart: he promise to bring
>   me where is *Anne Page*: by gar he deceiue me too.
>
> *Euan.* Well, I will smite his noddles: pray you follow.
>
>   (III.i.116-125)

This decision they reiterate later that same day when Evans tells Caius:

>   I pray you now remembrance to morrow on
>   the lowsie knaue, mine Host.
>
> *Cai.* Dat is good by gar, withall my heart.
>
> *Eua.* A lowsie knaue, to haue his gibes, and his
>   mockeries.   (III.iii.254-260)

And this is the last reference in the text to the revenge plan of Caius and Evans.

A rereading of the excerpt from IV.v. cited above will reveal that the lines that Caius and Evans utter when they warn the Host about the Cozen-Germans and the Duke who is not to come do not in any way give a hint that the horse-stealing episode is the revenge plan which the two duelists have executed. It is, of course, possible not to accept at face value the words of Evans and Caius when, after giving their news to the Host, they tell him respectively, " 'tis not conuenient you should be cozoned" and "I tell you for good will."

The interlinking, then, of the two incidents—the trick played on the Host and the revenge scheme—comes not on the authority of text but from the interpretations of editors seeking a resolution for the revenge scheme, which otherwise dies in III.iii. In finding the very characters who vow to get even with the Host furnishing him information about how he has been duped, the zealous editors—bent on creating a more intelligible text of the *Merry Wives*—have, in my estimation, succumbed to the obvious temptation. They have declared that the trick is the end result of an incompletely worked out and unspecified revenge plan.

But let us test this revenge–horse-stealing hypothesis from another standpoint. From study of the dueling episode it becomes clear that Caius and Evans are highly respected men in the community. One is the minister, the other a physician who numbers among his patients members of the Court. The matter which brings them into conflict is rather trifling. The Host, called in as intermediary, conceives a plan that is in no way harmful. It will simply let the two men carry out the semblance of a duel, give the choleric doctor an opportunity to "let off steam," and bring the matter to an end by stressing that each is too valuable a member of the community to risk his life in a foolish duel. Caius and Evans may feel a bit humiliated in the way the Host tricked them, with the result that their immediate reaction is to plan some sort of revenge. But to steal three horses—no doubt post horses—is a scheme that is completely incompatible with the characters of the doctor and parson as well as with Shakespeare's portrayal of village life in the play. With the value of horses so great in Elizabethan days, the theft of three would have been a major crime, far out of line with the degree of revenge

Caius and Evans might have planned. Note the usually cheerful Host's reaction at the opening of IV.vi.: "Master *Fenton*, talke not to mee, my minde is heauy; I will giue ouer all." And it is only Fenton's offer to make good the theft that brings him around:

> *Fen.* Yet hear me speake: assist me in my purpose,
> And (as I am a gentleman) ile giue thee
> A hundred pound in gold, more then your losse.
>
> (IV.vi.3-5)

Now it is entirely possible that in the interval between planning their revenge and the theft of the horses, Caius and Evans simply got over their grudge against the Host. In fact, whether consciously or otherwise, Shakespeare follows an Elizabethan theory of psychology in his portrait of Dr. Caius which bears out this conjecture. In a study of Caius as a choleric type of individual, John L. Stender calls attention to the fact that the Elizabethans recognized different forms of choler.[4] He cites, for example, W. Vaughan's *Directions for Health* in which Vaughan had divided choler into "open" and "hidden" categories. A man of the "open" choleric humour flew into sudden anger but soon got over it whereas one who was a "hidden" choleric had to transfer his anger into active revenge. Caius was of the "open" type. As support for placing him in this category, Stender observes that Caius releases Simple unharmed when he discovers him in his home (I.iv.70-119); that he quickly becomes reconciled with Evans when the Host reveals the trick at the conclusion of III.i.; and that (accepting the face reading) Caius warns the Host out of good will that no Duke de Jarmany is expected. If this theory be accepted, it makes a valid base for my contention that Shakespeare meant the Caius-Evans plan to terminate

4 "Master Doctor Caius," *Bulletin of the History of Medicine*, VIII (January 1940), 133-138.

when the two men had time to ruminate on the events that brought on the imbroglio.

Shortly after this incident, having somehow learned of the activities of horse-thieves in the neighborhood, Evans rushes to alert the Host. He realizes only too well that the Host is not above trickery and mockery on occasion. This, however, is not sufficient reason to make an innkeeper fall victim to horse thieves—especially when supplying post horses is an essential part of his livelihood. Caius also has a warning. But it is marked by a subtle difference. He does not repeat what Evans has said; instead he comments solely on the fact that no German duke is expected at court—a piece of information which, it has been established, he was able to furnish. What neither the parson nor the doctor realizes is that the warnings have come too late.

Perhaps Shakespeare, in his haste to complete the play, failed to round off this plot as he should have. Had he not similarly left the Shallow-Falstaff quarrel of the opening scene dangling? But to presume that the horse-stealing episode is founded on the Caius-Evans revenge scheme is to take unnecessary liberty with the text in my opinion.

If neither the Pistol-Nym nor Caius-Evans plans can serve to explain the presence of the horse-stealing episode in the play, what other possibility exists? W. W. Greg, who wrestled with various interpretations over the years, has leaned toward a theory that the episode was once more fully developed but is now so fragmentary because it may have contained some type of libelous material which had to be cut away.[5] Greg still retains this view in spite of a strong attack on it by E. K. Chambers. Chambers observed

---

[5] See 1910 Q1 facsimile, pp. xx-xxii, xlii; notes to lines 1344-71, p. 85; line 1530, p. 93; lines 1561-62, p. 93; line 1586, p. 94. Also *The Editorial Problem in Shakespeare*, p. 72 and *The Shakespeare First Folio*, pp. 336-337.

that if any censoring had taken place, the references to the Germans and a duke would have also been cut away.[6] Instead, Chambers suggested that the horse-stealing incident be considered as an independent entity which Shakespeare injected into his play but which, under the pressure of hasty composition, he could not work out and left in its fragmentary state.

But why should Shakespeare even have wanted to include such material in the *Merry Wives*? Professor John Crofts seems to have found the answer, although his fanciful reconstruction of Quarto-Folio relationship for the play[7] prevented him from recognizing it. Impressed by the tales of horse-stealing and posting scandals that had attached themselves to Count Mompelgard ever since Charles Knight had proposed him as the life model for the Duke de Jarmany, Crofts decided to make a thorough investigation of the subject.[8] His researches proved that although the Count had been issued a warrant to take up post horses without charge during his stay in England,[9] never had he nor any of his associates or representatives on subsequent missions become involved in a posting scandal. His appetite whetted by his study of contemporary records, Crofts —now feeling that the horse-stealing incident may have been rooted in some other posting scandal—pursued the matter further. He discovered two major scandals that occurred within months of the date Hotson had suggested for the composition of the *Merry Wives*.[10] The first, which involved Le Sieur Aymar de Chastes, the Governor of Dieppe, took place on September 4, 1596. The second, revolving around an illegally issued warrant by the Lords

6 *William Shakespeare*, I, 432.
7 *Shakespeare and the Post Horses*, pp. 131-140.
8 *ibid.*, pp. 11-18.
9 There is a copy in Rathgeb's *Badenfahrt*. See Rye, p. 47.
10 *Shakespeare and the Post Horses*, pp. 18-21, 32-43.

Thomas Howard and Montjoy, broke on November 17, 1597.

De Chastes, observes Crofts, was a distinguished political figure of the time. In holding Dieppe for Henry of Navarre, he had played an important role in securing the throne for the French monarch. De Chastes was also well known in English Court circles, having served as Envoy Extraordinary to England for Henry on several occasions. At the very time of the 1596 incident he was hurrying from England, where he had come as a member of the French ambassador's entourage for the ratification of the Treaty of Greenwich, to prepare for an English embassage on its way to Henry. On September 3, 1596, there was issued "An open placard to all her Majesty's publique officers to see Monsieur La Chatte, Governour of Deepe, provided of a convenient nomber of post horses and carriages for himself and retynew to Dover, &c."[11] This warrant, Crofts explains, gave no power to de Chastes either to requisition horses for himself or to take them beyond the stage for which they had been hired. But on September 4 he attempted both these things. The incident is described in Sir John Leveson's report to the Lord Chamberlain, sent from Gravesend on September 9, 1596:

"Has received answer from Dover, from Mr. Lieutenant of the castle there, touching 'the abuses offered to the governor of Dieppe at Gravesend and Rochester.' It appears that the governor complains that they could not obtain horses or carts at Gravesend, and received opprobrious words from the hacqueney men there, and that a certain woman dwelling in or near to the sign of the Horn took a gentleman of the governor's company by the beard with

[11] Dasent, XXVI; 136; also reproduced in Crofts, *Shakespeare and the Post Horses*, p. 150.

extreme violence, and had struck the governor himself had not a gentleman put her back.

"On receipt of this, repaired this morning to Gravesend and took examinations; which show that 'there were two horses in the stable of William Clarke of the Horn, which horses two gentlemen of the governor's company were desirous to have, and because they were the horses of strangers left there and no hacqueneys, they were locked up in a stable, the door whereof two Frenchmen did break open to take out the said horses, and the wife of William Clarke, whose husband was then out of the town, came into the stable and would have stayed the said horses there; and thereupon the Frenchmen thrust her from them and overthrew her, as she saith, and took out the said horses.' The wife denies that she pulled any by the beard; but says she 'was so amazed with the blow that one of the Frenchmen gave her, that she would have stricken him if she had found any staff or cudgel readily.' There are no witnesses, but one who saw the governor come out of the stable holding his hand on his beard 'as though one had been pulled by the beard.' As for the Rochester men, the horses which had been taken from Gravesend to Rochester being taken on to Sittingbourne and payment only made as far as Rochester, the hacqueneymen stayed the horses in the street there for the horsehire to Sittingbourne, and some disorder ensued. Has three or four of the men in custody, and asks what punishment he shall inflict upon the woman and them. Has forborne to send up the portreeve of Gravesend, for, the constable being sore sick, 'there would have been much disorder, and the Duke and his train could not have been accommodated of such horses, carriages, and other things as was fit.' "[12]

[12] *H.M.C. Salisbury*, VI, 375. Also reproduced in Crofts, p. 149. In commenting on these actions, Crofts notes that since the French posting system was very disorganized at this time, de Chastes was probably following the habits of experienced French travelers (p. 150, n. 12).

Knowledge of the entire affair must have been wide-spread. Aside from what word was spread by the actual participants, the Lieutenant of Dover made a report, the Governor of Upnor conducted an on-the-spot inquiry, the Lord Chamberlain received an account which he passed on to the Principal Secretary.[13] Crofts, therefore, sees Shakespeare seizing upon the incident and turning de Chastes and his two servants into the three German horse-thieves. Crofts (p. 20) concludes, "It is incredible that theatre-goers of 1597, searching their memory for the originals of the three German horse-thieves, would have been able to pass over this recent and notorious exploit by three French ones. . . ."

But having developed this theory, Crofts discards it in favor of one more in harmony with his untenable revisionist theory for the text of the *Merry Wives*. He believes that the de Chastes incident may have initially served as the inspiration for the horse-stealing subplot, but that when (according to Crofts' theory) Shakespeare revised the play in late 1597, he depicted the events of the second post-horse scandal—the Chard affair of November, 1597.[14] Even though he admits his hypothesis is highly conjectural, Crofts has labored much too hard to tailor the events of the Chard scandal to fit his theory of text, even dealing with matters that occurred after the *Merry Wives* presumably had been written. And he ignores the important question of whether "cosen garmombles" or "cozen Germans" was the original reading at IV.v.78-80.

<hr>

13 Crofts, p. 20.

14 This affair is an extremely complicated one—as Crofts develops it—having in its background a power struggle between Essex and the Lords Montjoy and Howard. In essence, Montjoy and Howard issued a post-warrant at Plymouth although they had no power to do so. One John Howard (or Heywood) presented it at Chard on November 17, 1597, and got away with receiving post horses. He was assigned a guide with whom he got into a fight. Legal action was taken which dragged on through 1599. See Crofts, pp. 32-36.

Let us, therefore, return to the cast-off de Chastes incident and pick up where Crofts left it. The mission which had brought de Chastes to England as an assistant to the Duke of Bouillon climaxed many months of Anglo-French negotiations for a mutual defense treaty against Spain. In international relations, the entire period from late 1595 through the summer of 1596 had been a trying one for England. In January, it had even appeared as if Henry IV of France might sever ties with his allies. Then in April came the fall of Calais—increasing the anxiety of the English. But by summer, with the victory at Cadiz and the promise of an Anglo-French treaty, new confidence swept over the English people. Undoubtedly the public was keenly aware of all that was happening. The treaty negotiations were probably closely followed.[15] By July 8, as a letter from Lord Burghley to his son Sir Robert Cecil indicates, the English were planning to dispatch a diplomatic mission to the French king.[16] Burghley also mentions in his letter that perhaps the occasion might be used "To carry the Garter to His Majesty" (the King having been elected to the Order in 1590). The end of the following month a French delegation, headed by the Duke of Bouillon, arrived in England, "came to the Court then at Greenwich, and there by her Maiesties oath confirmed the league of amity and peace betwixt the two Realms of England and France."[17] The week before the Duke arrived in England, the Earl of Shrewsbury had received his instructions to head the reciprocating mission to France. Not long after the Duke of Bouillon left England, Shrewsbury and his retinue departed on what had become a three-fold mission: (1) to obtain the ratification of the new treaty, (2) to invest Henry IV with

15 Many contemporary documents carry accounts of these negotiations. See, for example, B.M. Stowe MS. 132, fols. 114-135v.

16 *Cal. S.P. Dom., Eliz., 1595-1597,* p. 253.

17 Stow, p. 777.

the Garter, (3) to present Sir Anthony Mildmay as the resident ambassador to the French Court.[18] No other diplomatic mission to France during Elizabeth's reign was charged with three such major duties at the same time.[19]

It was to prepare the town of Dieppe for this embassage of the Earl of Shrewsbury that de Chastes had been hurrying home when he became involved in the posting scandal on September 4, 1596. The English ambassadors arrived at Dieppe on September 23, where, as Sir William Segar notes in his account of the mission,[20] they "were very nobly entertained and feasted, the first night by the Commander of Diepe, Mounsieur de Chate. . . ." Then they continued on to Rouen for the signing of the treaty and the investing of Henry. The entire period of the embassage was one of great splendor and festivity.[21] On October 30, the ambassadors and their retinue returned to England richly rewarded with jewels and money.[22]

Six months later Shakespeare presumably was writing *The Merry Wives of Windsor*. Knowing that he was preparing a play for a Garter function, he let his memory range over recent events touching on the Order. The only extraordinary incident was the investiture of Henry IV. That certainly was outstanding since it had been one of the three charges of the Earl of Shrewsbury during the mission to France the previous fall. Recalling that mission, which from its exceptional nature was still vivid in courtly circles, Shakespeare suddenly recollected the de Chastes

[18] Birch, *Memoirs of the Reign of Queen Elizabeth . . .* , II, 120.

[19] All Elizabethan diplomatic missions to France have been catalogued by F. J. Weaver, "Anglo-French Diplomatic Relations, 1558-1603," *Bulletin of the Institute of Historical Research*, IV (November 1926), 73-86; V (June 1927), 13-22; VI (June 1928), 1-9; VII (June 1929), 13-26. The mission of the Earl of Shrewsbury is described in VII, 23-24.

[20] B.M. Harl. MS. 1355, fol. 24.

[21] Stow, pp. 777-782.

[22] See B.M. Harl. MS. 1355, fol. 25 for rewards distributed by the Earl of Shrewsbury and King Henry.

posting scandal. He remembered that the French diplomat had been returning home at the time to arrange a fitting welcome for the Earl of Shrewsbury. Surely, Shakespeare pondered, this association could be worked into the play. But how? De Chastes was too distinguished a man to caricature openly. Besides, his trip to England was occasioned by major diplomatic negotiations. And the English entourage, in its turn, had returned home laden with rewards.

The fertile brain of the playwright toyed with the incident, and suddenly Shakespeare saw how he could take this de Chastes posting scandal and fictionalize it while still giving it topical significance. Shakespeare was aware that the Duke of Württemberg was to become a Knight of the Garter. He also knew that the Duke's election stemmed from political expediency, not merit. Furthermore, as a result of the Hanse troubles, Shakespeare saw only too plainly how unpopular the Germans were among the English at this time. It took only a flick of the creative pen to turn the three Frenchmen of the September 4 horse-stealing incident into three Germans; to make their leader a duke instead of a governor; and to change the locale from Gravesend and Rochester to Windsor and its environs. A hit was made at the Germans; the horse-stealing episode was retained; the Order of the Garter association was preserved. The only trouble is that having written the incident, Shakespeare was unable to integrate it into the main body of the play. He left it as a fragment, perhaps intending to rework it. But with so little time available, he never touched it again. And the horse-stealing episode joined the other bits of unfinished business in the *Merry Wives*.

The above reconstruction is admittedly conjectural. However, it is based on my conviction that the horse-stealing incident was conceived as an independent entity rather than as an extension of one of the revenge schemes in the play. Professor Crofts' discovery of the de Chastes posting scandal has been an invaluable contribution in recovering the origins of the episode. It is unfortunate that his theory of text led Crofts in other directions, never to inquire why there should have been a horse-stealing subplot or why Shakespeare should have turned Frenchmen into Germans.

The actual scheme by which the Germans tricked the Host into parting with his horses provides Shakespeare with another opportunity to reflect the spirit of Windsor at installation time. The Host has been forced to turn away all the guests from the Garter Inn in order to accommodate a party of Germans who are of the retinue of a German duke. Upon their arrival, the Germans request three horses. These the Host consents to give them, but is determined to make them pay plenty because they are to have his house "a week at command." When the Germans obtain the horses, off they ride, leaving Bardolph in a slough of mire. Instead of saucing his guests, it is the Host who is sauced. But is it not odd that so experienced an innkeeper as the Host should fall prey to a gang of horse thieves?

If the theft had occurred at any other time than when some grand affair—and particularly a Garter installation— were taking place in Windsor, one would indeed heap blame upon the Host for being so trusting with his horses. However, ever since the first foreign princes had been elected to the Garter in the late fourteenth century, Windsor inhabitants had become accustomed to receiving foreign rulers or their proxies when Garter installations were held. In Elizabeth's reign alone, before the election of the Duke of Württemberg, six other foreign potentates who had become members of the Order were installed at Windsor. Thus there is no reason why the Host at installation time should have been suspicious either of Germans or of the fact that they wanted horses to meet their master, a duke.[23] Nor should the Host have found it strange that

[23] The equanimity with which Windsor proprietors face sudden de-

only the retainers were lodged at his inn. The Garter records indicate that solely the Knights-elect, the commissioners of investiture, and immediate servants were allotted quarters in Windsor Castle.[24] "The rest," as Ashmole notes, "had their diet prepared in the Town at their own Inns."[25]

Aside from its fragmentary structure in terms of plot, the horse-stealing episode contains a few textual oddities. Most puzzling of these is the Folio reading "there is three Cozen-Iermans, that has cozened all the *Hosts* of *Readins*, of *Maidenhead*, of *Cole-brooke*, of horses and money. . ." (IV.v.78-80) which appears in the Quarto as "there is three sorts of cosen garmombles,/Is cosen all the Host of Maidenhead & Readings. . ." (ll. 1364-1365). Reference has already been made (Chapter VII) to critical opinion that *garmombles* represents a verbal scrambling of Garter and Mompelgard and therefore is to be interpreted as a slight satiric shaft aimed at the persistent Duke of Württemberg. If, as it seems, this was Shakespeare's intention, the *cosen garmombles* reading must have first appeared in the original text. When the actor who played the Host made his memorial reconstruction about 1601, he merely retained it.

---

scents of foreign dignitaries even today was amply demonstrated to me on April 15, 1957. While having lunch in a restaurant in town, I heard the proprietress receive a telephone reservation for tea for four for the High Commissioner of Ghana. She accepted the reservation as a matter of course with the comment to a waitress, "You know what to do."

24 Ashmole, p. 342.

25 Ashmole, p. 344. Ashmole was studying the installation of 1584. However, he indicates that this taking over of an inn ("at command") by the retinue of a Knight-elect was apparently a standard practice, one which was marked with a tangible testimonial of the occasion. Among the items a Knight-elect had to prepare for his installation, "Lastly, there hath been commonly provided a convenient number of Lodging Scutcheons, of the Elect-Knight's Arms, invironed with a Garter, with his Stile and Titles underneath; and these Garter [Garter King of Arms] also gets in readiness, for it hath been an ancient custom, to distribute at the Inns in the Knight's passage, to and at Windesor, these Scutcheons, to be set up in the principal Rooms of those Houses, as a memorial of the honor of the Knights Installation" (Ashmole, p. 336).

Since the phrase occurs in an exchange of dialogue between Evans and the Host, and is such an odd one, no great memory feat need be presumed in the Host's recalling and using it. Nor need we worry about the intelligibility of *garmombles* to a provincial audience. If anyone detected the topical allusion, well and good. If not, the word may have slipped by as a piece of nonsense or as Welsh mangling of the English language by Evans.

If the Folio text still contained the *garmombles* reading in 1601, when and why was the alteration made? It has previously been pointed out that any topical allusions to the Duke of Württemberg would have been recognized by a London, and especially a Court audience, at least through 1604. In this fact lies the clue to the substitution. In his study of the two variant readings, H. C. Hart proposed that the Folio alteration might be traceable to a changed relationship between Duke Frederick and the English Court once the Duke had been invested with the Garter in 1603.[26] This suggestion merits further probing.

Recalling our earlier discussion of Frederick's dealings with King James, it will be remembered that in July 1603, the Duke sent congratulatory letters to James as the new monarch. Later that month he dispatched Buwinckhausen to England to take up the matter of his Garter investiture. Buwinckhausen's meeting with the King was a cordial one. On September 18, James signed the certification of investiture, and sent Robert Lord Spencer of Wormleiton to head the investing delegation. Members of the delegation were splendidly entertained during their stay in Stuttgart and returned home laden with rewards. The following April Frederick sent Count Phillip von Eberstein as his proxy for the installation at Windsor. Nowhere in these dealings do we find the nagging Duke we encountered during Elizabeth's reign. And nowhere do we find Frederick a

[26] Arden edition, pp. lxxv-lxxvi.

necessary political pawn as a result of the Hanse affair. In fact, that unpleasant episode in Anglo-German relations had been resolved about the very time that Lord Spencer's delegation was on its way to Württemberg. Therefore, Hart's suggestion about the changed status of Frederick in James's Court is a sound one.

Shakespeare's company also gained new status under James. In May 1603, the troupe became the King's Men. As such nine members received the rank of Grooms of the Chamber and served at various state functions as royal retainers rather than as actors. In their new capacity as members of the Royal Household, they were aware of what was going on at Court, what foreign dignitaries were present, etc. For example, from August 8 to 27, 1604, twelve of the company attended the Spanish ambassador, Juan Fernandez de Velasco, during his stay at Somerset House.[27]

On November 4, 1604, the company was ordered to present the *Merry Wives* at Court. There were enough of the original shareholders in the King's Men who would have remembered the 1597 performance and the satire on the Duke of Württemberg in this play. They also undoubtedly were acquainted with the November 1603 investiture delegation to Frederick, perhaps had spoken with some of its members. They may even have served during the 1604 Garter Feast which Count von Eberstein attended. Therefore they could not have been ignorant of the changed status of Frederick in the English Court. Not wishing to risk offense at this Court performance, the company shareholders probably suggested to Shakespeare that he alter the only "delicate" phrase in the horse-stealing episode: *cosen garmombles.*

This is the type of procedure many critics think Shakespeare followed about this same time in cutting from II.ii

---

[27] Warrant for payment of services in P.R.O. AO 1/388/41.

of Q2 of *Hamlet* the references to Denmark as a prison (ll. 244-277) and the criticism of the children's companies (ll. 354-379) so as not to offend the new Queen, Anne of Denmark, who was also patroness of the Children of the Revels.

The change from *cosen garmombles* to *Cozen-Iermans* is just the sort that a facile writer would hit upon. *Cosen* (according to the OED) is an alternate spelling for both *cousin* and *cozen*, the latter having the dual meaning of "cousin" (the relative) and "to cheat, defraud by deceit." "Mon cousin," it will be remembered, is the mode of address employed by both Elizabeth and James in all their letters to Frederick. *Garmombles* is Mompelgard, who, of course, is a German prince. The process of reasoning Shakespeare engaged in is at once obvious. In substituting the term *Cozen-Iermans*, he has created a clever aural pun. Literally, as the capital letters prove, the term means "cheating or deceiving Germans"—a perfect description in terms of the plot. But there is also the word *cousin-german* (pl. *cousin-germans*) meaning the relative *cousin*, which in its figurative use returns us to the "term of address by a sovereign of another sovereign." The keener members of the audience at the 1604 Court performance could recognize the original allusion in its new guise while others would find a perfectly intelligible surface reading to the line. The change could have been made any time between May 1603 and November 1604, but it seems to be the type of alteration made specifically for a performance at which an unfavorable allusion might immediately be detected. That, of course, would have been the aforementioned production at Court. To hypothesize further, once the new reading had become part of the text, it was retained because Shakespeare (or his colleagues) realized that by late 1604 Mompelgard had become a stale subject for a topical allusion and that *Cozen-Iermans* was now a superior reading.

Even with the alteration of the sensitive phrase *cosen*

garmombles, the horse-stealing episode carries a slight anti-German tone to it—nothing unusual or outright offensive in a country where foreigners were looked on with suspicion. It is therefore interesting to find a line in the Folio text which may have been added to soften further the German references. Just after Bardolph informs him of the loss of the horses, the Host replies,"They are gone but to meete the Duke (villaine)/doe not say they be fled: *Germanes* are honest men" (IV.v.72-73). It is possible, however, that the Host paused slightly after "doe not say they be fled" and delivered the balance of the line as an afterthought—as if rationalizing away any doubt which may have flashed through his mind. The point can be pushed no further. But if *"Germanes* are honest men" represents an addition to the original text, as Dover Wilson suggests,[28] it must have been made at the same time as the *Cozen-Iermans* substitution.

There are other lines in this episode which have puzzled textual critics. At the beginning of IV.iii. Bardolph informs the Host,

> the Duke himselfe will be tomorrow at Court,
> and they [the Germans] are going to meet him.
>
> (ll. 2-3)

The Host replies,

> What Duke should that be comes so secretly?
> I heare not of him in the Court. . . .
>
> (ll. 4-5)

But from the balance of the scene it is obvious that the Host did expect the duke and his retinue. How to resolve this seeming contradiction?

If we think of the duke as Duke Frederick, we have a clue to the possible resolution of the problem. Contemporary documents prove—as previously cited examples in-

---

[28] New Cambridge edition, p. 127, note to lines 65-66. See also the Arden edition, p. lxxvi.

dicate—that all the efforts of Duke Frederick to obtain membership in the Order of the Garter before, during, and after the 1595 Breuning mission were undertaken in anything but a secret manner. Individuals both high and low—among them the Earl of Essex, Lord Burghley, and Sir Robert Cecil—were badgered by the Duke and his representatives to aid him in his desire. Where solicitation may have failed, bribery was resorted to. So widespread was this knowledge of Frederick's unabashed behavior that I believe Shakespeare, already planning his allusions to the Duke, commented on Frederick's conduct pointedly in the Host's lines "What Duke should that be comes so secretly?/I heare not of him in the Court. . . ." These lines, then, appear to be carefully written as laugh lines—delivered as an aside. Though lost to later generations, their meaning was perfectly intelligible to Elizabethan audiences—especially to that select first-night Garter audience.

The horse-stealing subplot, in summation, must be considered as an isolated entity in the play. The sole justification for its existence lies in Shakespeare's desire to make topical satire on the Duke of Württemberg. All elements in the incident—the initial inspiration behind it; the action, character motivation, and setting; even the textual alterations and the Host's comment about the duke's arriving in secret—can be traced to this satirical shaft aimed at Duke Frederick, the "Duke de Jarmany." That the horse-stealing subplot is an artistic failure in terms of the over-all structure of the *Merry Wives* is obvious. We can only understand Shakespeare's motives in inserting the episode in the script, and not be too harsh with him for his efforts to please his Garter audience with this topical material.

CHAPTER IX ✦ THE CHRONOLOGICAL RELATIONSHIP OF *The MERRY WIVES* TO THE *HENRY IV* AND *V* PLAYS

FOUR PLAYS relating the exploits of Falstaff and his roistering companions are: *1 Henry IV*, *2 Henry IV*, *The Merry Wives*, and *Henry V*. Early editors tried hard to establish a biographical link from play to play, and in their attempts they invented the most ingenious theories to trace the careers of whichever characters they found held over from one to another of these works. But there was always one play that completely thwarted them, for the *Merry Wives* simply refused to lend its characters to the establishment of any logical pattern. " 'It should be read,' says Johnson, 'between *King Henry IV*. and *King Henry V;*' 'no,' says Malone, 'it ought rather to be read between *The First* and *The Second Part of King Henry IV*.:' in good truth, 'it should be read between' none of them," wrote Alexander Dyce, "—being, as a story, complete in all its parts."[1]

The accuracy of Dyce's perceptive comment was not immediately recognized by his 19th century colleagues, but modern scholars have come to respect its validity. The *Merry Wives* is an island unto itself in the midst of the history plays. The six characters it shares with the *Henry* dramas are related in name only; otherwise they are entirely different creatures. This anomalous situation—unique in the writings of Shakespeare—can be explained in only one way. Shakespeare wrote the *Merry Wives* at command and with very little time to execute the finished play. Therefore (as the tradition records), since he was requested to take Falstaff from the *Henry IV* plays and

[1] *The Works of William Shakespeare* (London, 1857), I, clv.

now show him in love, Shakespeare simplified his task by borrowing at the same time the group of Falstaffian intimates and transplanting them to his new comedy. Such a statement calls for proof that *1* and *2 Henry IV* antedate the *Merry Wives*. And what of *Henry V*? If it closes the cycle, can we establish that Nym was created for the *Merry Wives* and then carried over into *Henry V*? By examining each of these plays individually, we shall see that the April 1597 date of the *Merry Wives* is indeed compatible with a sequence of composition which places *1 Henry IV* first, followed by *2 Henry IV*, the *Merry Wives*, and *Henry V*.

1 Henry IV was entered in the Stationers' Register on February 25, 1598, and published later that year by Andrew Wyse. Thus its terminal date is easily established. No documentary evidence exists for assigning its initial date, and scholarly opinion ranges from 1596 through 1597, with the majority of writers favoring the latter year. Until Leslie Hotson suggested the new dating for the *Merry Wives*, there was little need for scholars to attempt a closer dating for *1 Henry IV*. A generalized 1597 was more than sufficient. In fact, even Hotson was content to close the matter with the statement that "the two parts of *Henry IV*. must now be pushed back into the season 1596-1597."[2] But does *1 Henry IV* permit itself to be pushed around, or is it the victim of the brute force of a scholar seeking to make a theory valid?

Actually there is evidence for a slightly earlier dating. George Chalmers, in his *A Supplemental Apology for the Believers in the Shakespeare Papers* (London, 1799), pp. 326-330, found several passages in *1 Henry IV* that appear to reflect on topical events that occurred about the summer of 1596:

2 *Shakespeare versus Shallow*, p. 130.

(1)   So shaken as we are, so wan with care
        Finde we a time for frighted Peace to pant,
        And breath shortwinded accents of new broils
        To be commenc'd in Stronds a-farre remote

                                (I.i.1-4)

These lines, Chalmers believes, obliquely refer to the
Cadiz expedition of June 1596.

(2) Chalmers finds in Falstaff's remark, "the poor abuses
of the time, want countenance" (I.ii.174) a reflection of
Camden's comments about military preparations in 1596:
"There were a parcel of loose fellows, who went about the
kingdom, under the counterfeit authority of the *Queen's
Pursuivants*, with sham warrants, taking away, by force,
plate, jewels, and whatever they could find." A proclamation of May 3, 1596, bears on this observation of Camden's,
warning "against sundry abuses, practised by divers lewd
and audacious persons falsley naming themselves Messen-
gers of Her Majesty's Chamber, travelling from place to
place, with writings counterfeited, in form of warrants;
to the great slander of her Majesty's service, and abuse of
her loving subjects."

(3) The remark on the death of Robin the Ostler,
"Poore fellow neuer ioy'd since the price of oats rose, it
was the death of him" (II.i.13-14), Chalmers takes as an
allusion to the high cost of grain in 1596 and the Queen's
proclamation on July 31 "for the Dearthe of Corne."

Further topical significance has been found by J. E.
Morris.[3] Noting the abuses in impressment described in
the play (IV.ii.12-54), Morris calls attention to the efforts
of Sir John Smythe in 1596 on behalf of impressed men.
Smythe even tried to get some of the Essex-trained bands
to mutiny, and, as a result, was brought before the Star

[3] "The Date of 'Henry IV,' " TLS, January 28, 1926, p. 62.

Chamber in June 1596. Perhaps Smythe's efforts were not in vain, for a Council impressment order of September 12 makes note that the men receive good treatment.

The above-cited references are not to be taken as direct topical allusions. Like any good playwright, Shakespeare was influenced by the world around him. And in one form or another bits of that world found their way into his plays. Such topical passages are hard to ferret out. When discovered, they tempt their finder (possibly as compensation for his labors) to make definitive statements about their significance in establishing facts of Shakespeare's career. However, when a body of passages reflecting contemporary events that occur within a narrow time span is found in a play, then the date they point up merits serious consideration. The passages discussed by Chalmers and Morris have one thing in common: they call attention to happenings during the summer of 1596. Here then is probably the earliest possible date for *1 Henry IV*.

Evidence for the summer of 1596 as the outer limit for the initial date comes from another source: the substitution of the name *Sir John Falstaff* for *Sir John Oldcastle* in the *Henry IV* plays. It is now commonly accepted that the change was made at the request of one of the Lords Cobham, descendants of the historical Sir John Oldcastle, who took umbrage at the portrayal of their revered ancestor in *1 Henry IV*. The problem has been to determine which Lord Cobham made the protest, William Brooke or his son Henry. I believe the father, in his capacity as the powerful Lord Chamberlain, caused Shakespeare to make the alteration to *Falstaff*.[4] Cobham became Lord Chamberlain on August 8, 1596, and died on March 5, 1597. Accordingly, the composition of *1 Henry IV* becomes fixed within those limits.

4 See above, Chapter VI.

The shadow of another court personality hangs over a late 1596 date. Three prominent scholars—Levin L. Schücking, G. B. Harrison, and J. Dover Wilson—working independently, have found striking similarities between certain passages in *1 Henry IV* and aspects of the career of the Earl of Essex in 1596.[5] In no case do these writers claim that Shakespeare deliberately modeled a character in *1 Henry IV* after Essex. Rather, as Harrison states, "Shakespeare finds opportunities in the character of Hotspur for significant speeches and takes them, but he is writing a play and not an allegory of the times."[6] Yet Dover Wilson points out, "I find it impossible to doubt that Southampton's poet had Essex in mind while writing these historical plays or that they were written primarily for 'the judicious' of the Essex circle. . . .

"All four plays in the series have points of contact with Essex. Essex like Prince Hal had been a scape-grace in his early days; his intrigues with ladies at court were notorious; he was fond of low life and boon companions. In 1596, he suddenly like Prince Hal, became a reformed character and took for a time to devotion and pious exercises."[7]

And Harrison cites, among others, two instances of points of similarity between events in Essex's life in 1596 and passages in *1 Henry IV*. In Act I there is the dispute between King Henry and Hotspur over prisoners which compares with the ransom dispute between Elizabeth and Essex in September 1596 over the captives Essex took in the Cadiz expedition.[8] And in III.i.177-189, Worcester's

5 Schücking, "The Quarto of King Henry IV., Part II," TLS, September 25, 1930, p. 752; Harrison, "Shakespeare's Topical Significances: II. The Earl of Essex," TLS, November 20, 1930, p. 974; Wilson, *The Essential Shakespeare* (Cambridge, England, 1937), pp. 95-107. Each writer interprets the data about the Earl of Essex differently in determining its use for dating the *Henry* plays.

6 "Shakespeare's Topical Significances: II. The Earl of Essex," p. 974.

7 *The Essential Shakespeare*, p. 96.

8 We should note, however, that the dispute is reported in Holinshed's

analysis of defects in Percy's character resembles a similar analysis that Bacon made of Essex sometime in 1596-97 (on an occasion unspecified by Harrison). Again we have an emphasis on occurrences in late 1596 to early 1597.

Acceptance of the validity of topical allusions as evidence for dating frequently depends on the period of time that may elapse between the occurrence of the event and the appearance of the reference—either as a direct allusion or as an echo of an event transmuted by the writer's magic to serve artistic ends. All the references discussed above seem to have been employed before they had a chance to grow stale. It is true that they would have remained fresh well into 1597. But there is no need to suppose that Shakespeare waited longer than necessary to employ them. Now, if the topical references cited served as the only evidence for considering an earlier date for the play, our case might be tenuous. But from the documented terminal date of composition and from the general stylistic characteristics of *1 Henry IV* in terms of Shakespeare's development as a dramatist, we know the play falls within the 1596-97 time span. Therefore the topical references become supporting evidence. Since they fall within the same calendar period, their primary function is to set the initial date as late summer, 1596. Add the evidence of the appointment of Lord Cobham as Lord Chamberlain in August, with his death in March 1597, and the limits for the composition of *1 Henry IV* are pretty well set. Actually, there is little point of dispute in deciding whether the play could have been written in late 1596. Close dating merely demands that we move its composition back a few months from a date which at best has merely been estimated as probable for *1 Henry*

Chronicles (III, 520-522): thus Shakespeare may have relied exclusively on his source without having given the slightest thought to the Elizabeth-Essex controversy.

IV. The feasibility of such a move has been shown. Undoubtedly *1 Henry IV* was completed during the autumn of 1596 and was performed at Court during the Christmas-Shrovetide play season.

One final consideration remains. Did *1 Henry IV* start the cycle of the Falstaff plays, or could the *Merry Wives* have been the first? On the basis of the dating evidence just presented, *1 Henry IV* must have appeared several months before the *Merry Wives*. Furthermore, there is the stage tradition that the Queen was so well pleased with the character of Falstaff that she wanted to see him in love. To build a case upon a stage tradition may seem far less substantial than "the hill of flesh" who plays the central role in that tradition. But there it is, and the tradition must be reckoned with. There is still another fairly conclusive piece of evidence. In III.ii. of the *Merry Wives,* when the Host proposes Fenton as a husband for Anne Page, Master Page replies, "Not by my consent I promise you. The Gentleman is of no hauing, hee kept companie with the wilde Prince, and Pointz" (lines 72-74). The lines are completely meaningless in terms of the plot of the *Merry Wives* (all the more so if that play had come first), but the reference to Prince Hal and Poins seems to indicate that Shakespeare had details from the *Henry IV* plays still crowding his brain.[9]

The problem of close dating *2 Henry IV* to determine its chronological relationship to the *Merry Wives* is a complex one. If we adopt the formula suggested by Matthias

---

[9] There is also a reference to Hal in the Quarto of the *Merry Wives.* At lines 1522-23 Falstaff states, "Ile lay my life the mad Prince of *Wales/* Is stealing his fathers Deare." The whole passage is extremely corrupt and has no counterpart in the Folio. There seems to have been a complete memory breakdown by the pirate with a desperate attempt to fill in with any material. Therefore, for present purposes, the Q lines can be dismissed.

Shaaber,[10] the task becomes greatly simplified: "The date of this play is usually fixed with reference to that of *1 Henry IV*, most commentators agreeing that the one followed the other after a short interval. Accordingly, almost any scholar's opinion about the date of this play may be ascertained by adding a few months to his date for *1 Henry IV.* . . ." According to my reckoning, then, *2 Henry IV* would be placed around January to April 1597. This method of dating the play is, however, somewhat mechanical and would appear rather arbitrary unless, as will be shown, other evidence from *2 Henry IV* itself were forthcoming to corroborate the date.

The play's terminal date is easily established by the Stationers' Register entry on August 23, 1600, and by the publication the same year of the quarto edition. Outside of the fact that Part 2 had to follow Part 1 and that publication data establish the terminal date, there is only one piece of documentary evidence for narrowing the limits and that is the reference in V.ii. of *Every Man Out of his Humour* (1599) to Justice Silence. Even mention by Francis Meres of *"Henry IV"* in the *Palladis Tamia* (Stationers' Register, September 1598) is of no help, for there is absolutely no way of determining whether Meres meant Part 1, Part 2, or simply referred to both parts collectively.[11] Therefore, all dates assigned by scholars are conjec-

---

[10] *New Variorum Edition of Shakespeare: The Second Part of Henry the Fourth* (Philadelphia and London, 1940), p. 516.

[11] I believe that attempts to date *2 Henry IV* on the basis of whether quarto title page headings and Stationers' Register entries specify Part 1 or Part 2 can never be entirely successful. The interval in the publication of the two plays is too great. Besides, we have no way of knowing how the play records of the Chamberlain's Men read. Entries in extant documents (Henslowe's *Diary*, court reward lists, etc.) seem to indicate that the dramatic companies cared little about accurate rendering of their play titles. Furthermore, we cannot even tell why the Chamberlain's Men waited until 1600 to release *2 Henry IV* for the press. An interesting parallel in the looseness in designating this pair of plays appears in an article by the drama critic Brooks Atkinson. In discussing the skill of Boris Pasternak in translating Shakespeare's plays into Russian, Atkinson states, "A poet

tural. And the stumbling blocks placed in the path of a backward movement have been mixed from a mortar of opinion and speculation.[12] Nor are the inferences now made in support of an earlier date (prepared from these same raw materials) to be considered as anything more than tentative. At this point in studies in Shakespearean chronology no definitive statement can be forthcoming for *2 Henry IV*.

Textual study produces strong evidence for composition not long after *1 Henry IV*. The quarto of the second part retains a speech prefix to I.ii.137 labeled *Old*. Since this quarto is generally held to be printed from foul papers, Shakespeare apparently momentarily slipped back into using the name he had so laboriously extirpated from Part 1. With the passage of a greater time interval between the two parts, one doubts whether such a slip would have occurred. The apology in the epilogue—"Old-Castle dyed a Martyr, and this is not the man"—also seems to indicate that the enforced name change incident was still very much in Shakespeare's mind.

---

himself and a man of moral insight, Mr. Pasternak has recreated in Russian 'Hamlet,' 'Romeo and Juliet,' 'Antony and Cleopatra,' 'Othello,' '*Henry IV*,' 'King Lear,' and 'Macbeth' " (*New York Times*, September 14, 1958, Sec. 2, p. 1). Italics mine for *Henry IV*.

The same chaotic situation exists for the *Henry VI* plays. There is no quarto for *1 Henry VI*. Both the Stationers' Register entry and title page of Q1 of *2 Henry VI* refer to "The/First part of the Con-/tention betwixt the two famous Houses of Yorke/and Lancaster. . . ." Yet Q1 of *3 Henry VI* is entitled "The true Tragedie of Richard/Duke of Yorke. . ." with no acknowledgment that it represents the second part of *The Contention* . . . . And when Pavier entered these two works in the Stationers' Register in April, 1602, he called them "The first and Second parte of Henry the VJt." Yet we know he was in error. The matter is so confusing that even the Stationers' Register entry for the Folio printing of *1 Henry VI* reads "The thirde parte of Henry ye Sixt"—undoubtedly, as Greg notes (*The Shakespeare First Folio*, p. 60, n.1), a slip occasioned by confusion over the Pavier entry.

12 G. B. Harrison, for example, set the date as Lent, 1598, "for no better reason than that there are two jokes about Lent diet" ("Shakespeare's Topical Significances: II. The Earl of Essex," p. 974).

The points so far developed in this discussion have been made without reference to the almost certain appearance of the *Merry Wives* in April 1597. What I have tried to establish is not *was 2 Henry IV* written between January and April of that year, but *could* it have been written. With the answer in the affirmative, the date of the *Merry Wives* becomes a terminal point for *2 Henry IV*, and permits introduction of additional supporting data for fixing the composition of the latter play in early 1597. Of significance here is the fact that there are two characters in *2 Henry IV* who have no roles in Part 1 but who find their way to Windsor in the *Merry Wives*: Ancient Pistol, "the foule-mouth'dst Rogue in England"; and Justice Shallow, that "forked Radish, with a Head fantastically caru'd vpon it with a Knife." It seems inconceivable in terms of the over-all structure of *2 Henry IV* that Pistol and Shallow could have made their journeys in reverse—that is, from Windsor via the *Merry Wives* to the London and Gloucestershire of *2 Henry IV*. They had to appear first in *2 Henry IV* and then in the *Merry Wives*.

Another point for placing the chronicle history before the *Merry Wives* centers on the rejection of Falstaff. Critical opinion is virtually unanimous in citing the rejection as necessary in terms of the philosophical structure of the *Henry* histories. We may not like the way Shakespeare has treated our "fat rogue." But that is beside the point. The rejection is motivated. In fact, it falls upon a Falstaff who has degenerated considerably from what he was in Part 1. Is it not possible, therefore, that Shakespeare, as a result of his treatment of the character, realized that Falstaff was also dead dramatically by the time he penned the last lines of *2 Henry IV*? Or, as H. B. Charlton says, "Is it indeed Henry, or is it Shakespeare who rejects Falstaff?"[13] Given

---

[13] "Falstaff," *Shakespearian Comedy* (London, 1938), p. 196. This point of view that artistic as well as plot demands brought about the rejection

such a premise, we can understand why it mattered little to Shakespeare to debase Falstaff so completely in the *Merry Wives*. He was done with the character, the promise in the epilogue of *2 Henry IV* that "our humble Author will continue the Story (with Sir John in it)" being in the words of Charlton, "a pathetic hope . . . it may still be possible to save Sir John."[14] Then comes the request to show Falstaff in love. What does it matter now that Shakespeare robs him of his good name, Falstaff's great riches having all been spent. But once Shakespeare had written the *Merry Wives*, he passed final sentence on Falstaff and had no choice but to send him to Arthur's bosom. (A consolation prize at least for not having nestled in the bosoms of Mistress Page or Ford.) Perhaps Shakespeare even consciously commented on his treatment of the "fat kidneyed rascal" in this passage from the *Merry Wives*:

> I would all the world might be cozond, for
> I haue beene cozond and beaten too: if it
> should come to the eare of the Court, how
> I haue beene transformed; and how my
> transformation hath beene washd, and
> cudgeld, they would melt mee out of my fat
> drop by drop, and liquor Fishermens-boots
> with me: I warrant they would whip me with
> their fine wits, till I were as crest-falne
> as a dride-peare. (IV.v.95-103)

Thus Falstaff may be throwing a backward glance to the old days when he considered himself "fortune's steward." And that it is possible chronologically to look back from

---

of Falstaff, I arrived at independently before reading Professor Charlton's essay. Since his statement of the case is so apt, I develop the discussion no further.

[14] *Shakespearian Comedy*, p. 197.

the *Merry Wives* to *2 Henry IV* has been, I believe, amply demonstrated.[15]

In contrast to the complexity of close dating *2 Henry IV*, stands the case for *Henry V*. The entry in the Stationers' Register "to be staied" on August 4, 1600; Pavier's entry of August 14; and the issuance of Q1 this same year clearly establish the terminal date for *Henry V*. Its initial date is set by the allusion of the Chorus in Act Five (lines 22-34) to the Irish campaign of the Earl of Essex which commenced on March 27, 1599, and terminated disastrously the following September 28. Critical accord fixes the date of the play as the summer of 1599.

In determining the chronological relationship of the *Merry Wives* to the *Henry* plays, only one problem appears in regard to *Henry V*. In which work did Nym first figure, the *Merry Wives* or *Henry V*? E. K. Chambers believes it was in the latter play "because the description of Nym as 'corporal' [in the *Merry Wives*] would be meaningless, if

[15] Although I am inclined to the view that *2 Henry IV* was completed shortly before Shakespeare turned to the *Merry Wives*, I find of great interest a theory advanced by Henry N. Paul in the New Variorum edition of *1 Henry IV*, p. 355. Paul thinks that *1 Henry IV* was presented at Court during Shrovetide, 1597, and that Shakespeare immediately set to work on a sequel. However, he got only as far as IV.iii. of the second part when he was interrupted in order to write the *Merry Wives* at royal command for presentation shortly before the May Garter installation. He resumed work on *2 Henry IV* thereafter, and presented it the following winter. If Paul be correct, his hypothesis in no way invalidates the points I have made above. Both Pistol and Shallow have been introduced and their characters established before IV.iii; the degeneration of Falstaff has already been made evident; and surely by the time Shakespeare, or any dramatist, has reached almost the end of his play, he knows fairly well how he will compose his conclusion; thus the rejection would have been planned before Shakespeare interrupted his labors on the play. Indeed, if Shakespeare worked as Paul proposes, he would have become all the firmer in his resolution to drop Falstaff at the end of *2 Henry IV* once he had digested how completely he had debased the knight in the *Merry Wives*. What the Paul theory does—if I am not too liberal in interpreting it—is state that for all intents and purposes *2 Henry IV* was written before the *Merry Wives*, but performed after it. I in no way find such a conclusion incompatible with my own conjectures on this difficult problem of dating *2 Henry IV*.

he had not already made his appearance on the battle-
field."[16] Two lines in the comedy contain the description
Chambers refers to. In II.i.129 Pistol calls, "Away sir Corporall Nim"; and a few lines later (138) the cony-catcher
tells Ford, "My name is Corporall Nim." Of course there
is no military activity in the *Merry Wives* to warrant the
use of Nym's appellation "corporal" so that Chambers
seems to have valid grounds for placing the *Merry Wives*
after *Henry V*. But is it necessary to assume a literal cause
and effect sequence—that is, military campaign, hence title
equals *Henry V*—or is it possible to advance any other explanation which would make Nym a corporal without a
war in the earlier *Merry Wives*?

The title page of the 1602 *Merry Wives* Quarto supplies
a partial answer to the question in the phrase "With the
swaggering vaine of Auncient Pistoll, and Corporall Nym."
Ancient Pistol? Nowhere in the text is Pistol so designated.
As has been explained earlier, these lines of the title page
were undoubtedly added as an advertising stunt by the
publisher in order to capitalize on the popularity Nym
and Pistol had attained in *Henry V*; for the parts of the
two rascals in the *Merry Wives* alone are scarcely sufficient
to warrant citation of the characters on a title page. But if
the Quarto publisher was so careless in not giving an accurate description of Pistol in the *Merry Wives* (even if it
were for business purposes), may not Shakespeare have
been equally lax in making Nym a corporal in the play?

Pistol and Nym are but two of Falstaff's followers in the
Merry Wives. Bardolph is the third. Bardolph the "with-
ered serving man" turned tapster. He holds no military
rank in the *Merry Wives*, yet of the three cronies he is the
only one who fought through two campaigns under his
Captain Jack. The first time we meet Bardolph is in *1*

16 *William Shakespeare,* I, 434.

Henry IV. Again he is one of three who serve Falstaff, the other two being Gadshill and Peto. Gadshill drops out early in the play (II.iv). When Hal procures Falstaff the charge of foot in the campaign against the rebels, Falstaff makes Bardolph and Peto members of his company. Bardolph receives no rank, but Peto becomes a lieutenant (IV.ii.9). When Captain Jack takes the field in *2 Henry IV*, he again has three underlings: Bardolph, Peto, and Pistol. Pistol is his ancient (II.iv.74, 120, 165, 186);[17] Bardolph has been promoted to corporal (II.iv.167, III.ii. 235, 244); but alas, Shakespeare has failed to retain Peto's lieutenancy. In fact, Shakespeare compounds his inconsistency by assigning at one point the title of lieutenant to Pistol (V.v.94). Similar incongruities appear in *Henry V*. Although Sir John is no longer in it, Shakespeare retains the crew of three he gave the fat knight in the *Merry Wives*: Pistol, Nym, and Bardolph. Pistol is once again the ancient he was in *2 Henry IV*; Nym hangs on as corporal, and Bardolph is suddenly a lieutenant. But there are anachronisms in their titles within the play. At II.i.41 Bardolph addresses Pistol as "lieutenant." Dover Wilson tries to explain this slip as an attempt by Bardolph to placate Pistol.[18] Yet in the same breath Bardolph calls Nym "corporal," and Nym equally deserves to be placated at this point. Again, at III.ii.2 Nym addresses Bardolph as "corporal," but this is the same Nym who at II.i.2 says, "Good morrow Lieutenant *Bardolfe*."

This complete lack of consistency (which the chart below shows in summary) leads one to the conclusion that Shakespeare's designation of Nym as a corporal in the *Merry Wives* must be regarded as an anachronism. The

17 When Doll and Quickly refer to Pistol as "captain" in II. iv., they do so either out of flattery or pique. In his attempts to restore peace, Bardolph addresses Pistol by his correct title of ancient.
18 New Cambridge edition, p. 133, note to line 38.

fact that neither Pistol nor Bardolph bears a military title in this play only underscores the point. Shakespeare conceived of Falstaff's followers in groups of three. When he placed them in the history plays, he indiscriminately gave them military ranks which he neither remembered from play to play nor even within the same play.[19] Fresh from working on the two parts of *Henry IV*, Shakespeare slipped and made Nym a corporal. But the fact that he is a corporal in the *Merry Wives* cannot be considered as a cogent argument for dating the play after *Henry V*.

CHART OF MILITARY TITLES IN THE FALSTAFF PLAYS

(X indicates character does not appear)

	1 Henry IV	*2 Henry IV*	*Merry Wives*	*Henry V*
Falstaff	captain	captain	none	X
Bardolph	none	corporal	none	lieutenant, corporal
Peto	lieutenant	none	X	X
Gadshill	none	X	X	X
Pistol	X	ancient, lieutenant	none	ancient, lieutenant
Nym	X	X	corporal	corporal

Further evidence against placing the *Merry Wives* after *Henry V* lies in the aforementioned reference to the wild Prince and Poins at III.ii.72-74 of the *Merry Wives*. To suppose that after having just completed his testimonial to English monarchy with a portrayal of "the mirror of all Christian kings," Shakespeare would have so backtracked to make an unnecessary allusion to the wild prince if he

[19] Paul Jorgensen, in *Shakespeare's Military World* (Berkeley and Los Angeles, 1956), pp. 64-66, suggests, furthermore, that Shakespeare had little technical knowledge of army ranks. See also Karl Wentersdorf, "Shakespearean Chronology and the Metrical Tests," *Shakespeare-Studien: Festschrift für Heinrich Mutschmann* (Marburg, 1951), p. 173.

were writing the *Merry Wives* after 1599 is unthinkable. He even drops Poins from *Henry V* so that "the star of England" will bear no taint from his former associates. In marked contrast, the wild Prince-Poins reference does carry over the exact character grouping Shakespeare had been working with in both parts of *Henry IV*—the Falstaff and company, Hal and servant group of roisterers. It is entirely natural when laboring under the pressure of a command performance for Shakespeare to retain a vestige or two of these newly completed histories and to use the material, consciously or subconsciously, when needed.

The sequence of composition for the *Henry* plays in relation to the *Merry Wives*, accordingly, is proposed as follows:

1 Henry IV	Late 1596
2 Henry IV	January-April 1597 (at least completed if not performed)
Merry Wives	April 1597
Henry V	Summer 1599

Most problematical is the date assigned to *2 Henry IV*. Further probing into the chronology of both parts of *Henry IV* is, however, beyond the scope of this study. What I have proposed to show in this chapter is that there is no material in the *Henry* plays which may interdict a dating of the *Merry Wives* in April 1597 for performance at the Feast of the Garter.

CHAPTER X + A HYPOTHETICAL RE-CONSTRUCTION OF THE WRITING OF THE *MERRY WIVES*

"O FOR a Muse of fire, that would ascend/The brightest Heaven of Invention" and make known to the investigator working at a distance of four hundred years what really went through Shakespeare's mind as he took pen to hand and created the *Merry Wives*. Then would theory give way to certainty and the play reveal itself either truly or falsely as a Garter play. But since dramatic history has not been kind to us in supplying abundant facts about the genesis of the *Merry Wives*, we can but call upon our reason and imagination to interpret as objectively as is humanly possible the scanty data available. What concludes this study, therefore, is a summary of one writer's interpretation of such material. My remarks are predicated upon the word *suppose*. Suppose that Shakespeare wrote the *Merry Wives* especially for performance at the Feast of St. George in April 1597—as has been maintained throughout this book—then may not the evidence adduced for such a claim permit the reconstruction of the birth of the play in the manner now to be described?

The point of departure for making this reconstruction lies in establishing a meaningful association between the *Merry Wives* and the Order of the Garter. Again and again the Order acts like a magnet drawing the *Merry Wives* to it. Direct allusions—especially the carefully wrought paean to the Order in the fifth act—furnish the initial impetus for deducing an intimate relationship. These allusions lead us to inquire when during the period 1593-1602 the play could have been written for performance at a

Garter function. With a variety of contemporary events militating against a selection of any other year in this period, 1597 appears the probable choice, and indeed has strong evidence supporting it. Lord Hunsdon, Shakespeare's patron, was then elected to the Order. The Duke of Württemberg, the only German ruler made a Garter knight between 1579 and 1612, received his long-sought membership—a significant fact when we recollect the Duke de Jarmany of the play. And the Feast of St. George was celebrated that year in a splendid manner. At the Feast, St. George's Day itself furnished the most opportune time for presenting a special performance of the *Merry Wives*.

The events leading up to this performance have their beginning with a conjectured presentation of *1 Henry IV* at Court during the 1596-97 Christmas-Shrovetide play season. At this time Lord Hunsdon's Men acted before the courtiers on December 26 and 27, 1596; January 1 and 6, February 6 and 8, 1597. (Perhaps another of the six plays they performed was *2 Henry IV*.) Queen Elizabeth was captivated by the figure of Falstaff and commented, as tradition tells us, how much she would like to see the fat knight in love in a future play. In some manner her fancy reached the ears of her favorite cousin George Carey, Lord Hunsdon. Hunsdon, who had assumed the title only the previous July, made a mental note of her wish. As patron of Shakespeare's company he could do something to realize it should the proper occasion arise. And he found such an occasion not long after the Shrovetide play season had ended. On April 17 and 23, respectively, Hunsdon was appointed Lord Chamberlain and elected a Knight of the Garter. Twice the recipient of high honors within the same week, Hunsdon felt impelled to express his appreciation to the Queen in some special way. He recalled her desire to see a play depicting Falstaff in love, and promptly commissioned Shakespeare to write such a work for presentation on St. George's Day. While Hunsdon was not able to give Shakespeare too much advance notice, he could give him more than the fourteen days reported by Dennis. For Hunsdon knew at least three weeks before he officially received his new honors that they were to be conferred upon him. As a matter of fact, he had gradually been assuming more and more of the duties of the Lord Chamberlain's office before he was certified to the post. Through his position, therefore, he was able to call upon the various divisions of the royal household whose services might be needed to realize this specially commissioned court performance.

But a period of about three weeks was not much time for even a facile dramatist to turn out a play. And Shakespeare was further burdened by the request to depict Falstaff in a romantic entanglement. However, by 1597 Shakespeare had written about fifteen plays; he was no longer a novice, but a skilled craftsman as well as a literary artist. The craftsman side of him, accordingly, rose to the challenge. And, like so many master dramatic technicians in times of stress, he began to search for an old play which he could easily adapt for the occasion.

Now, no direct source has ever been found for the *Merry Wives*. Striking plot parallels do appear in the "Tale of the Two Lovers of Pisa" from Tarlton's *News Out of Purgatorie*; in "Of Two Brethren and their Wives" from Barnaby Riche's *Riche his Farewell to the Military Profession*; and in *Il Pecorone* (Day 1, Novella 2) of Ser Giovanni Fiorentino.[1] But the matter of positive identi-

[1] See Dorothy Hart Bruce, "*The Merry Wives* and *Two Brethren*," sp, xxxix (April 1942), 265-278; S. G. Thomas, "Source of 'The Merry Wives of Windsor,'" tls, October 11, 1947, p. 528; Kenneth Muir, *Shakespeare's Sources: Comedies and Tragedies* (London, 1957), I, 259-260; Geoffrey Bullough, ed. *Narrative and Dramatic Sources of Shakespeare* (London and New York, 1958), II, 4-8. Bullough reprints the above-mentioned tales on pp. 19-44.

fication is still unresolved. Furthermore, most students of the sources of the *Merry Wives* feel that even if these tales provide the raw material for the comedy, Shakespeare did not dramatize them directly.[2] Rather, in his race against time he seized upon an already created dramatization.

Fleay postulated that this old play was *The Jealous Comedy*, a work which he found entered in Henslowe's *Diary* for a performance by Lord Strange's company at the Rose on January 5, 1593.[3] Pollard and Wilson concurred with Fleay when they made their study of the *Merry Wives* many years later.[4] Their basis for claiming that *The Jealous Comedy* must have been the old play which Shakespeare revamped into the *Merry Wives* is predicated mainly on the observation that the subject matter of the *Merry Wives* also is jealousy. Pollard and Wilson even went one step further, and theorized that *The Jealous Comedy* must have depicted London middle-class society. Wilson and Quiller-Couch in the New Cambridge *Merry Wives* delved still more deeply into this suggestion, finding what they believed was supporting evidence for the middle-class aspects of the two plays in the *Merry Wives* Quarto.[5] From this version they concluded that the original of Dr. Caius must have been a London merchant. How else can one account for his closet being called a "counting house"? And why should Caius order Rugby to look out "ore de stall" unless a stall actually stood outside the house? Moreover, do not the lines in the fairy scene:

2 K. M. Lea points out in *Italian Popular Comedy* . . . (Oxford, 1934), II, 431-433, that although the stories have an older claim as possible sources for the *Merry Wives*, certain Commedia dell' Arte scenari exist which show affinity with the central plot of the *Merry Wives*.

3 *A Chronicle History of the Life and Work of William Shakespeare* . . . (London, 1886), pp. 210-212.

4 "The 'Stolne and Surreptitious' Shakespearian Texts: 'The Merry Wives of Windsor' (1602)," TLS, August 7, 1919, p. 420.

5 New Cambridge edition, pp. xxii-xxiii.

Where is *Pead?* go you & see where Brokers sleep,
And Foxe-eyed Seriants with their mase,
Go laie the Proctors in the street,
And pinch the lowsie Seriants face. . .

(Q, 1473-1476)

describe London?

The entire foundation for positing *The Jealous Comedy* as the old play which Shakespeare fashioned into the *Merry Wives* is built on the quicksand of a drama which has only a solitary reference in Henslowe's *Diary*. And as for the so-called London middle-class references, White has successfully demolished the attempt to make the original of Caius a merchant by noting that according to the OED, "counting house" in the sense of office was a legitimate use of the word in Elizabethan times. "Stall" for "stile" (the F reading), White further suggests, may be a compositor's error.[6] Even with the pure London of the above-quoted lines, first pointed out by H. C. Hart, the identification is completely fallacious since these lines must be read in conjunction with the preceding speech of Sir Hugh:

Come hither *Peane,* go to the countrie houses,
And when you finde a slut that lies a sleepe,
And all her dishes foule, and roome vnswept,
With your long nailes pinch her till she crie,
And sweare to mend her sluttish huswiferie.

(Q, 1467-1471)

Furthermore, these passages are found only in the corrupt (and later reconstructed) Quarto. The case for linking the *Merry Wives* and *The Jealous Comedy*, therefore, must be regarded as tenuous.[7]

A more satisfactory attempt to determine the nature of the *Ur-Merry Wives* has been made by Oscar James Camp-

---

[6] "Textual History," pp. 146-147.
[7] See also John Munro, "Some Matters Shakespearean—I," TLS, September 13, 1947, p. 472.

bell.[8] Campbell believes that the original play must have been constructed along the lines of conventional Italian comedy and that the prototype of Falstaff had been a pedant or a scholar.[9] But the unknown author of this old play took the stock Italianate situation of the clever lover (a role often assumed by the pedant) who deceives the husband and reversed it so that the duped one became the lover. This farce situation he grafted on to the traditional Italian tale of young lovers forced to outwit objecting parents. The love story itself followed the usual line of an *amorosa* wooed by three suitors. The girl, of course, had her maidservant who acted as a go-between.

As Shakespeare set about to transform this dual-plot *Ur-Merry Wives*, he initially appears to have attempted to create a new plot situation. For in the opening scene of the play he momentarily recaptures the flavor of the Gloucestershire country scenes of *2 Henry IV*. Possibly he felt that inasmuch as he was obliged to employ Falstaff in his new play, he might ease his task of creating additional characters by also carrying over the fat knight's cronies from the script he had just been working on. But after completing about one hundred-odd lines, inspiration failed him. Then, instead of extirpating the scene and starting over, in his haste to complete the *Merry Wives*, he simply left it dangling, unintegrated—a mute witness to the relentlessly ticking clock at his back.

Professor Campbell believes that Shakespeare, thus thwarted in his efforts to compose this new material, next tried to incorporate the Falstaff crew directly into the

8 "The Italianate Background of *The Merry Wives of Windsor*," *Essays and Studies in English and Comparative Literature* (Ann Arbor, 1932), VIII, 81-117. I have, in the main, summarized this account.

9 At the time of the writing of the above article Professor Campbell conjectured that the original play was *The Jealous Comedy*. He has now tempered this opinion in a letter to me dated September 15, 1958, and currently believes that we can go no further than to state that the old play was written in the Italian mode.

Italianate comedy he was reworking. When he realized that he could not match them all with characters in the ur-play, Shakespeare made what use he could of them and then wrote them out of the script. Only Dame Quickly was retained, for it required little effort to turn her into the needed go-between.

Falstaff, of course, became the duped lover. However (turning from Campbell), here another problem arose—one long ago pinpointed by Malone. Even though the Queen had wanted to see Falstaff in love, "Shakespeare," as Malone so aptly states, "knew what the queen, if the story be true, seems not to have known . . . Falstaff could not love, but by ceasing to be Falstaff. He could only counterfeit love, and his professions could be prompted, not by the hope of pleasure, but of money."[10] Still, as another astute critic—Professor Parrott—points out, there are traces of the old Falstaff in the play, primarily in I.i and III.iii, which serve as an indication that Shakespeare had tried hard to create Falstaff in his former image before the demands of the Italianate model forced him to transform the fat knight into the duped scholar-lover.[11]

Thus far Shakespeare's task was not complicated. Once he had decided to utilize the dual plot of the old play and to transform the Falstaffian contingent, he could proceed in a fairly mechanical manner. Renaming characters, adapting incidents—these were routine for a skilled craftsman. But if Shakespeare had stopped here all he would have had was a delightful comedy in the popular Italian

[10] *The Plays and Poems of William Shakspeare*, I, Part ii, 308-309 n.
[11] *Shakespearean Comedy* (New York, 1949), pp. 257-261. Parrott notes, "In action Falstaff is almost always the butt and dupe, a very different character from the domineering Falstaff of *King Henry IV Part II*. In speech, on the other hand, barring the few exceptions noted when he seems to fall out of his role, he constantly recalls the Falstaff of the histories. It is hardly possible to account for this inconsistency except on the hypothesis that Shakespeare rewrote hastily and under pressure an old play in which some form of the Tarlton story was presented in dramatic form" (pp. 260-261).

mode. His commission would have been executed; the Queen's desires satisfied.

Yet Shakespeare was aware of the occasion for which Lord Hunsdon had requested the writing of the play. He also knew full well the significance of April 23. Hunsdon's commission probably went no further than to specify that Shakespeare compose a drama showing Falstaff in love. But the dramatist, as he planned his work, apparently conceived the idea of incorporating into the script additional material that would serve as a tribute to the Order of the Garter. Thus not only would the *Merry Wives* fulfill the wish of the Queen, but it would also salute Shakespeare's patron on receiving membership in the Order. In other words, *The Merry Wives of Windsor* was to be Shakespeare's Garter play.

How easy to accomplish this. First alter the setting of the old Italianate play to Windsor. That very stroke sets up an immediate Garter association for an Elizabethan audience as we have previously seen in Chapter One. Make the time contemporaneous. The stock characters of the Italianate model readily become the middle-class village folk. Against the activities of their daily lives, with which the plot of the *Merry Wives* is primarily concerned, place an installation at the castle—the "grand affair" which the town (and castle) physician historically attended. And, to clinch the Garter-Windsor link, insert a tribute to the Order. Thinking along these lines, in the same scene with his tribute, Shakespeare adds a graceful compliment to the Queen who he knew would be in the audience for the initial performance. Imagine the impact of all this on Lord Hunsdon and the other Knights-elect who, seeing the play immediately upon their election, would have unfolded before their eyes the Windsor of a month hence—a Windsor in which some of the inhabitants went about their daily routines while others prepared for the installation cere-

monies which these Knights-elect would shortly be par-
ticipating in.

As Shakespeare works up the Garter aspects of his play,
he sifts his memory for any recent events involving the
Order which he may be able to exploit for dramatic ends.
He recalls the de Chastes posting scandal of the previous
September. De Chastes had been hurrying back to France
at the time to prepare a welcome for the Earl of Shrewsbury who was expected shortly to invest Henry IV with the
Garter. Good material here, notes Shakespeare, but not in
its present form. De Chastes cannot openly be caricatured.
Then Shakespeare remembers that among those to be made
Knights of the Garter is the Duke of Württemberg. It had
been common knowledge in courtly circles for several years
that the Duke badly wanted the Garter. And it was no
secret that he was finally getting it only to keep him happy
at a time when Elizabeth needed allies among the German
princes. A little ridicule of the German Duke would bring
smiles to the lips of a Garter audience. And a barb or two
at Germans in general would sit well with the spectators
in April 1597, since the Hanse troubles had undoubtedly
made the stranger-hating Englishmen look upon Germans
with even more than usual dislike. So Shakespeare meta-
morphoses the de Chastes horse-stealing incident into one
built around an *absentia* German duke. But writing
against the clock as he is, Shakespeare simply cannot integrate his horse-stealing subplot into the play proper. He
leaves it, isolated, just as he had done with the opening
scene—another monument in the play to the tyranny of
time.

In addition to experiencing difficulties with plot which
the exigencies of writing under pressure did not permit
him to solve successfully, Shakespeare ran up against one
minor crisis just before the *Merry Wives* received its proposed 1597 performance. Artistic considerations had led

him to select *Brooke* as the alias for *Ford*. In rehearsal, however, someone—perhaps Lord Hunsdon checking on the script—realized that *Brooke* also was the name of the lately deceased Lord Cobham. This was the same individual who had raised the protest over the use of the name *Oldcastle* in *1 Henry IV*. As if recollection of this incident alone might not have suggested prudence lest disrespect to the memory of the powerful lord have been suspected, further cause for restraint in the use of the name *Brooke* was furnished by the knowledge that Lord Cobham had been one of the most influential members of the Order. In 1595 he even had the distinction of serving as Lieutenant during the Feast of St. George. And here the *Merry Wives* was being readied for presentation before a Garter audience. Thus out of deference to the deceased William Brooke, Lord Cobham, Shakespeare altered Ford's assumed name from *Brooke* to *Broome*. A craftsman-like step, even if it left a botched pun in the text.

The *Merry Wives*, then, appears mainly the work of a highly skilled craftsman. But Shakespeare, although he was writing under pressure and with strict limitations as to the nature of his central plot, was also an artist. As such he was able to escape the shackles of the mere technician by casting the play in the mold of the new humours comedy. The spread in popularity of humours-type characterizations in literature had grown steadily from about 1592 until by 1596 such portrayals were in vogue with the writers. Not only with the writers, but with the sophisticated Londoners who, in the mid-1590's, made humours psychology the fad of the day—even carrying it so far as to interlard their conversation with the word *humour* at the slightest opportunity.[12]

As soon as Shakespeare sensed that the latest dramatic

---

12 G. B. Harrison, *Elizabethan Plays and Players* (London, 1940), pp. 154-156.

trend was toward humours portrayals, he turned to writing this type of comedy. Undoubtedly the success of George Chapman's *The Blind Beggar of Alexandria*—the first representative of the humours fad in the drama—influenced Shakespeare to try his hand at the new medium. *The Blind Beggar of Alexandria* opened at the Rose on February 12, 1596, and played twenty-two performances within the space of fourteen months, a record for its time. No other work among the seventy-seven presented by the Admiral's Men during the years 1594-97 was played so repeatedly in so short an interval.[13] In ascertaining the influence of the *Blind Beggar* on Shakespeare, one must not overlook the fact that nine performances of the play (November 6 and 12, December 2, 10, and 23, 1596; January 15 and 25, March 14, April 1, 1597) were given within the period that the *Henry IV* plays and the *Merry Wives* were probably written. As Shakespeare worked on these scripts, he had before him a constant living reminder of the success that could be achieved by incorporating elements of the new humours techniques in his scripts. And in the tiring room and tavern Shakespeare and his colleagues must have talked of the money Henslowe was making with that Chapman work. Perhaps the theatrical grapevine had even leaked word that Chapman was readying a second and more polished "comodey of vmers" for Henslowe.

This new play, *An Humorous Day's Mirth*, had its initial performance on May 11, 1597—almost three weeks after the postulated first production of the *Merry Wives*. True, there happen to be many likenesses between *An Humorous Day's Mirth* and Shakespeare's comedy.[14] Both plays deal with the intense jealousy of a husband for his wife. The

---

[13] See *Henslowe's Diary*, Chart D, II, 339-341.

[14] Though I reached this conclusion before reading William Bracy's book, I wish to acknowledge the more detailed study of the similarities on pp. 118-119 of *History and Transmission*.

two works present a gallery of humours characters—some from the same stockpile. There are Ford and Count Labervele, the jealous husbands; Slender and Labesha, the gulls. Also, as Bracy points out, Nym and Pistol resemble Blanuel in the way all three use peculiar phraseology. The plays are further alike in their overemployment of the word *humour*. (It occurs twenty-six times in the *Merry Wives*, a greater frequency than in any other play Shakespeare wrote from the start of his career through *Henry V* —the last relevant play for this study.) Yet it is impossible to think that one of these works served as the model for the other. The time element between their respective appearances—April 23 and May 11—is too narrow for such a hazardous conjecture. The similarities in makeup argue, rather, for the thesis that during 1596-97 a vogue for humours plays established itself upon the London stage, and that the playwrights were catering to the public taste.

Attuned to what he knew would be popular with his audience, Shakespeare again merged the artist and craftsman within him to produce two characters in the *Merry Wives* who stand among the most delightful and original creations in Shakespeare's comic gallery—Sir Hugh Evans and Dr. Caius. The Italianate original probably dictated the need for these two comic characters.

As Professor Campbell indicates,[15] Shakespeare could not completely transform the pedant of the old play into a Falstaff-like character. Therefore he needed another pedant to take up the slack—most apparent in the oral examination scene of IV.i. So Shakespeare invented Sir Hugh. In making him a Welsh parson, Shakespeare capitalized on the new popularity of the stage Welshman. W. J. Lawrence has shown that in the 1590's Welsh portraiture

---

[15] "The Italianate Background of *The Merry Wives of Windsor*," pp. 97, 109-110.

suddenly became very popular with theatergoers.[16] The craze started in 1593 with Peele's *Famous Chronicle of King Edward the First* and became established by 1595 with Drayton's *The Welshman* and Munday's *John a Kent and John a Cumber* (possibly a revision of *The Wise Man of West Chester*). Welshmen, Lawrence indicates, "were never assailed with the rancorous hostility so often a note of Irish and Scottish portraiture. . . . Such was the esteem for the Welsh that they were never considered outlanders." Shakespeare, with his usual sensitivity to new dramatic trends, was quick to exploit the possibilities of Welsh portrayals, advancing from a mere sketch of a captain in *Richard II* to the Welsh scene of *1 Henry IV* (with its forever-lost Welsh dialogue). Lawrence continues to trace the vogue right through to 1603, pointing out en route that Shakespeare's supreme contribution to the Welsh gallery is Fluellen. In creating a Welsh parson for the *Merry Wives*, then, Shakespeare was possibly thinking of the successful reception of his Glendower and daughter scene in *1 Henry IV* and responding to what he knew would be a well-received stage figure in 1597.

A similar consideration may have led to the birth of Dr. Caius. Following the Italianate model, Campbell notes,[17] Shakespeare required a third lover for his *amorosa*. Traditionally, at least one of the three *amorosos* was a grotesque type. And among the stock characters of Italian comedy can be found a "farced medico" type of pedant. Campbell infers that this was the third *amoroso* of the *Ur-Merry Wives*. But since Shakespeare had already reworked the pedant, he changed the character to a straight physician. Yet Shakespeare never comments on Caius as a physician. It is primarily as a choleric Frenchman that he is

---

[16] "Welsh Portraiture in Elizabethan Drama," TLS, November 9, 1922, p. 724.
[17] "The Italianate Background of *The Merry Wives of Windsor*," pp. 99-112, *passim*.

satirized.[18] The reason for making him French, Campbell suggests, is that "foreign physicians, particularly Frenchmen, were held in absurdly high esteem, especially by highborn and fashionable Englishmen. Caius boasts of the earls, knights, lords, and gentlemen whom he can number among his patients."[19] Hence Shakespeare may have picked his character to reflect another current fad in his only play of contemporary English life. By making Caius a humours character at the same time as a foreign physician, Shakespeare may have been twitting his high-born and fashionable first-night audience for following both the humours psychology and medical fads.

Shakespeare's task—as hypothetically reconstructed—was now completed. The *Merry Wives* had been written as ordered although Falstaff did not quite emerge as a romantic lover. Shakespeare had even gone beyond his commission and had composed a gay comedy which at the same time honored the august body before which the play received its initial performance on the evening of April 23, 1597. Thus Shakespeare's Garter play stands as vivid testimony to the playwright's ability to write on command, with little time, and for a specific occasion.

Five years later a version of the play, memorially reconstructed without authorization for acting in the provinces, found its way into print—albeit in a somewhat mangled form. While we have no way of knowing the fortunes of the play between its initial production and its publication in 1602, we can surmise that the *Merry Wives* achieved instant popularity once it was added to the repertory of the Lord Chamberlain's Men. If the play had met with failure, why should the actor-pirate have taken the trouble to make a reconstruction? Also, why should a publisher

---

[18] For a discussion of contemporary medical theories about the choleric humour of Frenchmen, see Stender, pp. 133-138.

[19] *Shakespeare's Satire* (New York, 1943), p. 80.

have gone to the expense of issuing a printed (and inferi-
or) version of what had already proved to be a "flop" on
the stage?

The play apparently retained its popularity into the
reign of James I. Both the Court performance of 1604 and
the reprint of the Quarto in 1619 attest to the continuing
interest in the *Merry Wives*. Thus far, however, only the
corrupt version of the play was available to readers. But
finally the more authentic text, which with the passage of
time had undergone but slight alterations (primarily in
the areas of topicality and excision of oaths), appeared
in the 1623 Folio.

Imperfect in its structure *The Merry Wives of Windsor*
is. But its flaws are those which are more apparent in the
study than on the stage. And as a stage work it has endeared
itself to countless audiences since the Knights of the Garter
first witnessed it in 1597. *Honi soit qui mal y pense.*

APPENDIX ✦ ABSENCE OF THE
MERRY WIVES FROM MERES'S LIST

ON September 7, 1598, Francis Meres's *Palladis Tamia*: *Wits Treasury* was entered in the Stationers' Register. Probably the most famous passage from this work is the oft-quoted

> As *Plautus* and *Seneca* are accounted
> the best for Comedy and Tragedy among
> the Latines: so *Shakespeare* among ye
> English is the most excellent in both
> kinds for the stage; for Comedy, witnes
> his *Gentlemen of Verona*, his *Errors*,
> his *Loue labors lost*, his *Loue labours*
> *wonne*, his *Midsummers night dreame*, &
> his *Merchant of Venice*: for Tragedy
> his *Richard the 2. Richard the 3.*
> *Henry the 4, King Iohn, Titus Andronicus*
> and his *Romeo* and *Iuliet*.

Meres makes no mention of *The Merry Wives of Windsor*. His failure to cite the play may at first glance appear an indictment of the theory of dating originally advanced by Hotson and now more fully developed in the present study. But we have not been chasing chimeras. Circumstantial evidence, which I have developed in considerable detail, exists to back up the thesis that the *Merry Wives* was written in 1597—at least a year before the Meres work appeared.

Rather than the theory of dating herein propounded, it is the reliability of Meres that should be questioned. Scholars, thanks largely to the research of D. C. Allen,[1] now

[1] *Francis Meres's Treatise "Poetrie,"* University of Illinois Studies in Language and Literature, XVI (1933), pp. 9-60; introduction to *Palladis*

recognize that Meres was not a particularly outstanding, trustworthy, or original critic. His work was mainly derivative, and Allen proves this by a detailed analysis of the *Palladis Tamia*, showing that Meres wrote primarily in the *imitatio* method taught in the schools of the time. As a result, most of the similes and quotations in *Palladis Tamia* can be traced to Erasmus's *Parabolae sive Similiae*. Historical references and anecdotes about Greek and Latin writers may be found in Ravisius Textor's *Officina*. Information about early English writers seems to be drawn from the works of Ascham, Puttenham, and Webbe. Allen also notes that Meres, without acknowledgment, has incorporated into the *Palladis Tamia* extracts from the prose romances of Greene, Warner, and Lyly. All these observations lead Allen to declare that "even [Meres's] historical data may be now questioned with justice."[2]

Aside from the great doses of imitation in Meres's writing, the *Palladis Tamia* exhibits another characteristic which should throw the gravest doubts on accepting Meres as an authority. The man is obsessed with symmetry. From the very opening passage of the section entitled "A Comparative Discourse of our English Poets, with the Greeke, Latine, and Italian Poets," he balances three against three, one against one, eight against eight, etc. In practically every paragraph of this section, and of the following two on musicians and painters, he makes an exact matching of the number of references, whether they be citations of works, authors, or genres. Furthermore, this desire for symmetry seems to have made Meres omit or alter material at will, even if he placed a writer in a wrong category as a result.

His methodology can be illustrated by comparing two listings from "A Comparative Discourse. . ." with two simi-

---

Tamia [1598] (New York, Scholars' Facsimiles & Reprints, 1938), pp. vii-viii.

2 *Francis Meres's Treatise "Poetrie,"* p. 60.

lar listings that appear in Stow's *Annals*. Meres opens his citation of poets with: "As Greece had three Poets of great antiquity, *Orpheus, Linus* and *Musaeus*; and *Italy*, other three auncient Poets, *Liuius Andronicus, Ennius & Plautus*; so hath England three auncient Poets, *Chaucer, Gower* and *Lydgate*." In the *Annals* (p. 811) the passage reads: ". . . the chiefe of our ancient Poets, recommended vnto vs, by worthy writers, are *Fryer* Bacon, Thomas Wickliffe, *sir* Ieoffry Chaucer *Knight*, John Gower *Esquier*, Iohn Lidgate *monke* of *Bury, doctor* Skelton, *sir* Thomas More, *Lord Chancellor*, master Iasper Heyward, *Gentleman*."

In another passage, Meres states: "As these Neoterickes *Iouianus Pontanus Politianus, Marullus Tarchaniota*, the two *Strozae* the father and the son, *Palingenius, Mantuanus, Philelphus, Quintianus Stoa* and *Germanus Brixius* haue obtained renown and good place among the auncient Latine Poets: so also these English men being Latine Poets, *Gualter Haddon, Nicholas Car, Gabriel Haruey, Christopher Ocland, Thomas Newtown* with his *Leyland, Thomas Watson, Thomas Campion, Brunswerd & Willey*, haue attained good report and honorable aduancement in the Latin Empyre." Stow (p. 812) renders this: "These following were Latine Poets. Gualter Hadon *Gentleman, Master* Nicholas Carre *Gentle. Ma.* Christopher Otland *Gent.* Mathew Gwyn *Doctor of Physick*, Thomas Lodge *Doctor of Physicke, Mast.* Tho. Watson *Gentle.* Thomas Campion *doctor of Physicke*, Richard Lateware *Doctor of diuinity* Master Brunswerd *Gentleman, Master doctor* Haruie, and *master* Willey *Gentleman*."

It appears as if the Meres passages may have provided the inspiration and model for placing a catalogue of prominent English writers in Stow's *Annals*. What is significant is that if he were influenced by Meres, Stow paid absolutely no attention to the symmetry in the Meres listings—

three for three in the first passage above and nine for nine in the second.[3] Thus, stylistic considerations aside, Stow cites eight of the older English poets and eleven of the neo-Latin. Assuming the other possibility of pure coincidence in these similar entries and even allowing for individual choice in cataloguing categories of writers, the *Annals* entries show us that more candidates were available than Meres chose to use. Therefore we can hardly avoid any other conclusion than that Meres manipulated his listings until they fit preconceived designs of symmetry.

Let us now take a second look at the passage with the list of Shakespeare's plays. It starts with a one to one ratio: Plautus for Latin comedy and Seneca for tragedy. Why the slight to Terence? Obviously to balance the citation of Seneca as *the* exponent of Latin tragedy. Next Meres is careful to note that Shakespeare is the best in both dramatic genres in England; hence the ratio remains constant. Now comes the list of plays. And here Meres balances six comedies against six tragedies. In his "tragedy" category he even fails to discriminate between "tragedy" and "chronicle history" (although here there may be some question as to what degree the Elizabethans considered chronicle history an independent dramatic genre). If we further relied on Meres, it would appear as if the twelve plays cited represented Shakespeare's total output at the time of publication of *Palladis Tamia*. This we know is

---

[3] I believe that Meres counted "the two Strozae" entry as one. Unfortunately, I was not able to consult all editions of the *Annals* on this matter of influences in compiling lists so similar in format. There is a strong probability that Edmund Howes added the lists in his continuation of Stow's *Annals* since both the 1615 and 1631 editions carry the quoted passages; the 1592 edition does not.

[4] Scholars who date *2 Henry IV* with a blanket 1598 completely hedge the problem. At least E. K. Chambers recognized that "Meres' notice in the autumn of 1598 might cover either one or two parts. . . ." (*William Shakespeare*, I, 383).

completely false. Yet if Meres were our sole witness, we should have no record of the existence of *1 Henry VI, 2 Henry VI, 3 Henry VI,* or *The Taming of the Shrew*—all written before 1598. Moreover (as has earlier been noted in Chapter IX), the entry for *Henry IV* does not clarify whether Meres meant Part 1, Part 2, or a collective heading for both parts.[4] Was this ambiguity also intentional for the sake of symmetry?

Thus Meres's list must be regarded only as a guide to Shakespeare's output through September 1598, not as a definitive catalogue of the dramatist's works as of that date. The omission of the *Merry Wives* from the *Palladis Tamia* account, therefore, cannot be interpreted as sufficient grounds for denying a pre-1598 date of composition for the play.

BIBLIOGRAPHY

A. EDITIONS OF THE PLAYS
OF WILLIAM SHAKESPEARE CONSULTED

1. INDIVIDUAL EDITIONS OF
The Merry Wives of Windsor

Daniel, P. A. Introduction to William Griggs, *Shakespeare's Merry Wives of Windsor: The First Quarto, 1602,* A Facsimile in Photo-Lithography, rev. London, 1888.
Greg, W. W., Introduction. *The Merry Wives of Windsor: 1602.* Shakespeare Quarto Facsimiles No. 3. Oxford, n.d. (Originally published by the Shakespeare Association, London, 1939.)
———, ed. *Shakespeare's Merry Wives of Windsor: 1602.* Oxford, 1910.
Halliwell-Phillipps, J. O., ed. *The First Sketch of Shakespeare's Merry Wives of Windsor.* London, 1842.
Hart, H. C., ed. *The Merry Wives of Windsor.* The Arden Shakespeare. London, 1904.
Wilson, J. Dover and Arthur Quiller-Couch, eds. *The Merry Wives of Windsor.* Cambridge, England, 1921.

2. INDIVIDUAL EDITIONS OF OTHER PLAYS

Evans, G. Blakemore, ed. *Supplement to Henry IV, Part I: A New Variorum Edition of Shakespeare,* in SQ, VII (Summer 1956).
Greg, W. W., ed. *Henry the Fifth: 1600.* Shakespeare Quarto Facsimiles No. 9. Oxford, 1957.
Hemingway, Samuel B., ed. *A New Variorum Edition of Shakespeare: Henry the Fourth Part I.* Philadelphia, 1936.
Shaaber, M. A., ed. *The First Part of King Henry the Fourth.* The Pelican Shakespeare. Baltimore, 1957.
———. *A New Variorum Edition of Shakespeare: The Second Part of Henry the Fourth.* Philadelphia and London, 1940.
Walter, J. H., ed. *King Henry V.* The [New] Arden Shakespeare. Cambridge, Mass., 1954.
Wilson, J. Dover, ed. *The First Part of the History of Henry IV.* Cambridge, England, 1946.

Wilson, J. Dover, ed. *King Henry V*. Cambridge, England, 1947.

――――. *The Second Part of the History of Henry IV*. Cambridge, England, 1946.

Wilson, J. Dover and Arthur Quiller-Couch, eds. *The Two Gentlemen of Verona*. Cambridge, England, 1921 (rev. 1955).

――――. *The Winter's Tale*. Cambridge, England, 1931.

3. COLLECTED WORKS

Capell, Edward, ed. *Mr. William Shakespeare his Comedies, Histories, and Tragedies*. 10 vols. London, 1767-68.

Clark, William George and William Aldis Wright, eds. *The Complete Works of William Shakespeare*. The Globe Shakespeare (1866, rev. 1911). New York, n.d.

Dyce, Alexander, ed. *The Works of William Shakespeare*. 6 vols. London, 1857.

Johnson, Samuel, ed. *The Plays of William Shakespeare*. 8 vols. London, 1765.

Knight, Charles, ed. *The Pictorial Edition of the Works of Shakspere*. 7 vols. London, [1839-42].

Kökeritz, Helge and Charles Tyler Prouty, eds. *Mr. William Shakespeares Comedies, Histories, & Tragedies*. A Facsimile Edition of the First Folio. New Haven, Conn., 1954.

Malone, Edmund, ed. *The Plays and Poems of William Shakspeare*. 10 vols. London, 1790.

Pope, Alexander, ed. *The Works of Shakespear*. 6 vols. London, 1725.

Rowe, Nicholas, ed. *The Works of Mr. William Shakespear*. 6 vols. London, 1709.

Theobald, L., ed. *The Works of Shakespeare*. 7 vols. London, 1733.

B. SELECTED BIBLIOGRAPHY OF PRIMARY SOURCES

1. MANUSCRIPTS

B.M. MSS. Add. 6298; 10,110; 12,508; 12,512.

――――. Cott. Galba D. XIII; Galba E. I; Titus C. X; Vesp. F. III.

――――. Harl. 304; 1355; 1760; 1776; 6064; 6834.

――――. Lansd. 76; 79.

――――. Stowe 132.

Blue Book (Liber Coeruleus). [Annals of the Order of the Garter from the commencement of the reign of Queen Mary I (1553) through April 23, 1621.]

Bodl. Ashm. MSS. 1108; 1109; 1110; 1111; 1113; 1115; 1125; 1131; 1134.

Coll. Arms MS. Collections Relating to the Order of the Garter, MS. Saec. XVI.

―――. Garter: Miscellaneous. (This is a collection of unclassified MSS in the custody of Garter Principal King of Arms.)

P.R.O. MSS. AO 1/386/35; AO 1/387/36; E 101/432/12; E 351/543; E 351/1795; E 404/132; E 404/133; LC 4/193; LC 5/182; SP 81/7; SP 81/8; SP 81/9; SP 82; Signet Office Docquet Books.

2. PRINTED WORKS

Anstis, J., ed. *The Register of the Most Noble Order of the Garter, From its Cover in Black Velvet, usually called The Black Book.* . . . 2 vols. London, 1724.

Arber, Edward, ed. *A Transcript of the Registers of the Company of Stationers of London, 1554-1640.* 5 vols. London, 1875-94.

Assum, Joannes Augustinus. *Panegyrici tres Anglowirttembergici.* . . . Tübingen, 1604.

Barret, Robert. *The Theorike and Practike of Moderne Warres, Discoursed in Dialoguewise.* London, 1598.

Birch, Thomas, ed. *The Court and Times of James the First; containing a series of Historical and Confidential Letters.* . . . 2 vols. London, 1849.

―――. *Memoirs of the Reign of Queen Elizabeth, From the Year 1581 till her Death . . . From the Original Papers of Anthony Bacon . . . And other Manuscripts never before published.* 2 vols. London, 1754.

Braun, George, and Francis Hohenberg. *Civitates Orbis Terrarum.* 6 vols. Brussels, 1572-1617.

Breuning von Buchenbach, Hans Jakob. "My most submissive narration of all that happened from the day on which I was graciously dispatched to England by the Court till my return, and of all that I have with the utmost sedulousness accomplished," in *Queen Elizabeth and Some Foreigners. . .* , ed. Victor von Klarwill, trans. T. H. Nash, pp. 357-423. London, 1928.

Breuning von Buchenbach, Hans Jakob. *Relation über seine Sendung nach England im Jahr 1595*. Stuttgart, 1865.

Calendar of State Papers, Domestic Series, of the Reign of Elizabeth, 1595-1597, ed. Mary Anne Everett Green. London, 1869.

――――. *1598-1601*, ed. Mary Anne Everett Green. London, 1869.

Calendar of State Papers, Domestic Series, of the Reigns of Elizabeth and James I., Addenda, 1580-1625, ed. Mary Anne Everett Green. London, 1872.

Calendar of State Papers, Domestic Series, of the Reign of James I., 1603-1610, ed. Mary Anne Everett Green. London, 1857.

Camden, William. *Britannia siue florentissimorum regnorum Angliae, Scotiae, Hiberniae chorographica descriptio* [1586, *rev.* 1607], 4th ed., trans. Edmund Gibson. 2 vols. London, 1772.

――――. *The History of the most Renowned and Victorious Princess Elizabeth. . .* [1615], 3rd ed. (1675), trans. T. Hearne. London, 1717.

Cellius, Erhardus. *Eques Auratus Anglo-Wirtembergicus*. Tübingen, 1605.

Chamberlain, John. *Letters Written by John Chamberlain during the Reign of Queen Elizabeth*, ed. Sarah Williams. London, 1861.

Chambers, E. K. and W. W. Greg, eds. "Dramatic Records of the City of London. The Remembrancia," *Collections*, Part I, pp. 43-100. The Malone Society Reprints. Oxford, 1911.

Chapman, George. *The Blind Beggar of Alexandria* (1598), ed. W. W. Greg. Malone Society Reprints. Oxford, 1928.

――――. *An Humorous Day's Mirth* (1599), eds. W. W. Greg and David Nichol Smith. Malone Society Reprints. Oxford, 1938.

Collins, Arthur, ed. *Letters and Memorials of State . . . Faithfully transcribed from the Originals at Penshurst Place. . . .* 2 vols. London, 1746.

Cunningham, Peter, ed. *Extracts from the Accounts of the Revels at Court in the Reigns of Queen Elizabeth and King James I*. London, 1842.

Dasent, John Roche, ed. *Acts of the Privy Council of England*, n.s., 1542-1604. 32 vols. London, 1890-1907.

De Lafontaine, Henry Cart, ed. *The King's Musick: A Transcript of Records Relating to Music and Musicians (1460-1700)*. London, [1909].

Dennis, John. *The Comical Gallant, or the Amours of Sir John Falstaffe*. London, 1702.

———. *Original Letters, Familiar, Moral and Critical*. 2 vols. London, 1721.

D'Ewes, Simonds, ed. *The Journals of all the Parliaments during the Reign of Queen Elizabeth, both of the House of Lords and House of Commons*. London, 1682.

Drayton, Michael. *Poly-Olbion*, IV in *The Works of Michael Drayton*, ed. J. William Hebel. Oxford, 1933.

Feuillerat, Albert, ed. *Documents Relating to the Office of the Revels in the Time of Queen Elizabeth*. London, 1908.

Gildon, C. "Remarks on the Plays of Shakespear," Supplement to *The Works of Mr. William Shakespear*, VII, ed. Nicholas Rowe. London, 1710.

Greg, W. W., ed. *Henslowe's Diary*. 2 vols. London, 1904 and 1908.

———. *Henslowe Papers: Being Documents Supplementary to Henslowe's Diary*. London, 1907.

Hawarde, John. *Les Reportes del Cases in Camera Stellata 1593 to 1609*, ed. William Paley Baildon. n.p., 1894.

Hentzner, Paul. "Extracts from Paul Hentzner's Travels in England," in *England as Seen by Foreigners. . .*, trans. and ed. W. B. Rye. London, 1865.

Heylyn, Peter. *The Historie of that most famous Saint and Souldier . . . St. George . . . The Institution of the most Noble Order of S. George, named the Garter. . .*, 2nd ed. London, 1633.

Historical Manuscripts Commission. *Calendar of the Manuscripts of the Most Hon. The Marquis of Salisbury Preserved at Hatfield House*. 18 vols. London, H.M. Stationery Office, 1883-1940.

———. *Report on Manuscripts in Various Collections*. II. London, H.M. Stationery Office, 1903.

———. *Report on the Manuscripts of Lord de L'Isle & Dudley*

Preserved at Penshurst Place. 4 vols. London, H.M. Stationery Office, 1925-42.

————. *Second Report of the Royal Commission on Historical Manuscripts.* London, H.M. Stationery Office, 1871.

————. *Sixth Report of the Royal Commission on Historical Manuscripts.* London, H.M. Stationery Office, 1877.

Holinshed, R. *Chronicles of England, Scotland and Ireland* [1587]. 6 vols. London, 1807-1808.

Klarwill, Victor von, ed. *Queen Elizabeth and Some Foreigners: Being a series of hitherto unpublished letters from the archives of the Hapsburg family*, trans. T. H. Nash. London, 1928.

Meres, Francis. *Palladis Tamia, Wits Treasvry: Being the Second part of Wits Commonwealth.* Modern Language Association Collection of Photographic Facsimiles. No. 32. London, 1598.

Murdin, William, ed. *A Collection of State Papers Relating to Affairs in the Reign of Queen Elizabeth from the Year 1571 to 1596. . . .* London, 1759.

Nichols, John, ed. *The Progresses and Public Processions of Queen Elizabeth. . . ,* new ed. 3 vols. London, 1823.

Old Cheque-Book, or Book of Remembrance, of the Chapel Royal, from 1561 to 1744, The, ed. Edward F. Rimbault. Camden Society Pub. New Series III. [Westminster], 1872.

Peck, Francis, ed. *Desiderata Curiosa: or a Collection of Divers Pieces Relating chiefly to Matters of English History . . . from the Original . . . MS. Copies . . . or . . . MS. Collections . . . .* 2 vols. London, 1732, 1735.

Peele, George. *The Honour of the Garter.* London, 1593.

Rathgeb, Jacob. *Kurtze und Warhafte Beschreibung der Badenfahrt. . . .* Tübingen, 1602.

————. "A True and Faithful Narrative of the *Bathing-Excursion,* which . . . Frederick, Duke of Wirtemberg . . . made to . . . England," in *England as Seen by Foreigners. . . ,* trans. and ed. W. B. Rye, pp. 1-53. London, 1865.

Rye, W. B., ed. *England as Seen by Foreigners in the Days of Elizabeth and James the First. . . .* London, 1865.

Rymer, Thomas. *Foedera [Syllabus (in English) of the Documents Relating to England and other Kingdoms . . . Known*

As "Rymer's Foedera"], trans. and ed. Thomas Duffus Hardy.
3 vols. London, 1869-85.

"The Statutes and Ordinances of the most Noble Ordre of
Saint *George*, named the *Gartier*, Reformed, explained, declared, and renewed by the moost High, moste excellent, and
mooste puissant Prince Henry the viii. . . ," in *Memoirs of
St. George . . . and of the Most Noble Order of the Garter*
by Thomas Dawson. London, 1714.

Stow, John. *Annales, or a General Chronicle of England: Be-
gun by John Stow: Continued and Augmented . . . by Edmund Howes*. London, 1631 ed.

Teshe, William. Untitled poem beginning "Within a Place or
Pallace richlye dight," in Nicholas Harris Nicolas, *History
of the Orders of Knighthood of the British Empire. . . ,* II,
Appendix H, xxix-xxxviii. London, 1841.

Wilbraham, Sir Roger. *The Journal of Sir Roger Wilbraham
. . . For the Years 1593-1616. . . ,* ed. Harold Spencer Scott.
The Camden Miscellany, x. London, 1902.

C. SELECTED BIBLIOGRAPHY OF
SECONDARY SOURCES

Albright, Evelyn May. *Dramatic Publication in England, 1580-
1640*. New York, 1927.

Allen, Don Cameron, ed. *Francis Meres's Treatise "Poetrie."*
University of Illinois Studies in Language and Literature,
XVI. Urbana, 1933.

———. Introduction, *Palladis Tamia* by Francis Meres [1598].
New York, Scholars' Facsimiles & Reprints, 1938.

Ashmole, Elias. *The Institution, Laws & Ceremonies of the
Most Noble Order of the Garter*. London, 1672.

Bald, R. C. " 'Assembled' Texts," *Lib.*, 4th Ser., XII (1931), 243-
248.

———, ed. *A Game of Chess*, by Thomas Middleton. Cambridge, England, 1929.

Baldwin, T. W. *The Organization and Personnel of the Shakespearean Company*. Princeton, 1927.

———. "William Kemp not Falstaff," MLR, XXVI (April 1931),
170-172.

Bartlett, John. *A Complete Concordance of Shakespeare*. New
York, 1953.

Baskervill, Charles Read. "English Elements in Jonson's Early Comedy," *Bulletin of the University of Texas,* Humanistic Series, No. 12, Studies in English, No. 1 (April 8, 1911).

Beltz, George Frederick. *Memorials of the Most Noble Order of the Garter.* London, 1841.

Black, J. B. *The Reign of Elizabeth 1558-1603.* Oxford, 1936.

Bowers, Fredson. "The Problem of Variant Forme in a Facsimile Edition," *Lib.,* 5th Ser., vii (December 1952), 262-272.

Boyce, Benjamin. *The Theophrastan Character in England to 1642.* Cambridge, Mass., 1947.

Bracy, William. *The Merry Wives of Windsor: The History and Transmission of Shakespeare's Text.* The University of Missouri Studies, xxv. Columbia, Mo., 1952.

Bradley, A. C. "The Rejection of Falstaff," *Oxford Lectures on Poetry,* 2nd ed., pp. 247-273. London, 1926.

Brock, Elizabeth. "Shakespeare's *The Merry Wives of Windsor*: A History of the Text from 1623 through 1821." 2 vols. Unpublished dissertation. University of Virginia, 1956.

Bruce, Dorothy Hart. "*The Merry Wives* and *Two Brethren,*" sp, xxxix (April 1942), 265-278.

Bruce, J. Douglas. "Two Notes on 'The Merry Wives of Windsor,'" mlr, vii (April 1912), 239-241.

Bullough, Geoffrey, ed. *Narrative and Dramatic Sources of Shakespeare.* ii. London and New York, 1958.

Buswell, John. *An Historical Account of the Knights of the most Noble Order of the Garter. . . .* London, 1757.

Cain, H. Edward. "Further Light on the Relation of *1* and *2 Henry IV,*" sq, iii (January 1952), 21-38.

Campbell, Oscar James. "The Italianate Background of *The Merry Wives of Windsor,*" *Essays and Studies in English and Comparative Literature,* pp. 81-117. University of Michigan Publications in Language and Literature, viii. Ann Arbor, 1932.

————. A letter to the author, dated at Columbia University, September 15, 1958, on the subject of *The Jealous Comedy* as a possible source for *The Merry Wives.*

————. *Shakespeare's Satire.* London and New York, 1943.

Chalmers, George. "The Chronology of Shakespeare's dramas," *A Supplemental Apology for the Believers in the Shakespeare Papers: Being a Reply to Mr. Malone's Answer, which was early announced, but never published.* . . , pp. 266-494. London, 1799.

Chambers, E. K. "Court performances before Queen Elizabeth," MLR, II (1906), 1-13.

———. "The Disintegration of Shakespeare," *Aspects of Shakespeare*, pp. 23-48. British Academy Lecture, 1924. Oxford, 1933.

———. *The Elizabethan Stage.* 4 vols. Oxford, 1923.

———. *Notes on the History of the Revels Office under the Tudors.* London, 1906.

———. *Sir Henry Lee.* Oxford, 1936.

———. *Sources for a Biography of Shakespeare.* Oxford, 1946.

———. *William Shakespeare: A Study of Facts and Problems.* 2 vols. Oxford, 1930.

Charlton, H. B. "Falstaff," *Shakespearian Comedy*, pp. 161-207. London, 1938.

Cheyney, Edward P. *A History of England: From the Defeat of the Armada to the Death of Elizabeth.* 2 vols. London and New York, 1926.

Collier, J. Payne. *The History of English Dramatic Poetry . . . and Annals of the Stage. . . .* 3 vols. London, 1831.

Cowl, R. P. "Some Literary Allusions in 'Henry the Fourth,'" TLS, March 26, 1925, p. 222.

Craig, Hardin. *A New Look at Shakespeare's Quartos.* Stanford, 1961.

Crofts, John. *Shakespeare and the Post Horses.* Bristol, 1937.

Cutts, John P. A letter to the author, dated at the University of Missouri, January 8, 1958, on the subject of music for court plays.

Dawson, Thomas. *Memoirs of St. George The English Patron; and of the Most Noble Order of the Garter.* London, 1714.

Dowden, Edward. *Shakspere.* London, 1877.

Draper, John W. "The Date of *Henry IV*," *Neophilologus*, XXXVIII (n.d.), 41-44.

———. "The Humor of Corporal Nym," SAB, XIII (July 1938), 131-138.

———. *The Humors and Shakespeare's Characters.* Durham, N.C., 1945.

Ehrenberg, Richard. *Hamburg und England im Zeitalter der Königin Elisabeth.* Jena, 1896.

Fellowes, Edmund H. *The Knights of the Garter 1348-1939* . . . . London, [1939].

Fleay, Frederick Gard. *A Chronicle History of the Life and Work of William Shakespeare: Player, Poet and Playmaker.* London, 1886.

――――. *A Chronicle History of the London Stage 1559-1642.* London, 1890.

"Friedrich I., Herzog von Würtemberg," by P. Stälin. *Allgemeine Deutsche Biographie.*

Gardiner, Samuel R. *History of England from the Accession of James I. to the Outbreak of the Civil War 1603-1642.* 10 vols. New Impression. London, 1899-1900.

Gaspey, Thomas. *The Life and Times of the Good Lord Cobham.* 2 vols. London, 1843.

Gilbert, Allan. *"The Merry Wives of Windsor"* in *The Principles and Practice of Criticism.* Detroit, 1959.

Gildersleeve, Virginia Crocheron. *Government Regulation of the Elizabethan Drama.* New York, 1908.

Gray, Henry David. "The Rôles of William Kemp," MLR, XXV (July 1930), 261-273.

――――. "Shakespeare and Will Kemp: A Rejoinder," MLR, XXVI (1931), 172-174.

Greer, C. A. "An Actor-Reporter in the 'Merry Wives of Windsor,'" N&Q, CCI (May 1956), 192-194.

Greg, W. W. *The Editorial Problem in Shakespeare: A Survey of the Foundations of the Text,* 2nd ed. Oxford, 1951.

――――. Letter of reply to Evelyn May Albright in "Notes and Observations," RES, IV (April 1928), 202-204.

――――. Review of *The Merry Wives of Windsor: The History and Transmission of Shakespeare's Text,* by William Bracy. SQ, IV (January 1953), 77-79.

――――. *The Shakespeare First Folio: Its Bibliographical and Textual History.* Oxford, 1955.

――――. *Some Aspects and Problems of London Publishing Between 1550 and 1650.* Oxford, 1956.

――――. *Two Elizabethan Stage Abridgements: The Battle of Alcazar & Orlando Furioso: An Essay in Critical Bibliography.* Oxford, 1923.

Haller, Eleanor Jean. "The Realism of the Merry Wives," *West*

Virginia University Studies: III. Philological Papers (Volume 2), (May 1937), pp. 32-38.

Halliwell-Phillipps, J. O. *An Account of the only Known Manuscript of Shakespeare's Plays, comprising Some Important Variations and Corrections in the Merry Wives of Windsor. . . .* London, 1843.

————. *Outlines of the Life of Shakespeare*, 2nd ed. London, 1882.

Hannigan, John E. "Shakespeare *Versus* Shallow," SAB, VII (October 1932), 174-182.

Harries, Frederick J. *Shakespeare and the Welsh.* London, 1919.

Harrison, G. B. *Elizabethan Plays and Players.* London, 1940.

————. *Shakespeare at Work: 1592-1603.* London, 1933.

————. "Shakespeare's Topical Significances," TLS, November 13, 1930, p. 939 (I. *"King John"*); November 20, 1930, p. 974 (II. "The Earl of Essex").

Hart, Alfred. *Shakespeare and the Homilies: And Other Pieces of Research into the Elizabethan Drama.* Melbourne, 1934.

————. *Stolne and Surreptitious Copies: A Comparative Study of Shakespeare's Bad Quartos.* Melbourne and London, 1942.

Harwood, T. Eustace. *Windsor Old and New.* London, 1929.

Herford, C. H. and Percy Simpson, eds. *Ben Jonson.* 11 vols. Oxford, 1925-52.

Hinman, Charlton. "Cast-off Copy for the First Folio of Shakespeare," SQ, VI (Summer 1955), 259-273.

Hope, W. H. St. John. *Windsor Castle: An Architectural History.* 2 vols. London, 1913.

Hoppe, Harry R. "Borrowings from *Romeo & Juliet* in the 'Bad' Quarto of *The Merry Wives of Windsor*," RES, XX (April 1944), 156-158.

Horne, David H. *The Life and Minor Works of George Peele.* New Haven, Conn., 1952.

Hotson, Leslie. "A Great Shakespeare Discovery," *Atlantic Monthly*, CXLVIII (October 1931), 419-436.

————. *Shakespeare versus Shallow.* Boston and London, 1931.

Jorgensen, Paul A. *Shakespeare's Military World.* Berkeley and Los Angeles, 1956.

Kirschbaum, Leo. "An Hypothesis concerning the Origin of the Bad Quartos," PMLA, LX (September 1945), 697-715.

Kirschbaum, Leo. *Shakespeare and the Stationers.* Columbus, Ohio, 1955.

Kreider, Paul V. *Elizabethan Comic Character Conventions: As Revealed in the Comedies of George Chapman.* University of Michigan Publications in Language and Literature, XVII. Ann Arbor, 1935.

Lawrence, W. J. "Assembled Texts in the First Folio," TLS, January 12, 1922, p. 28.

————. "Welsh Portraiture in Elizabethan Drama," TLS, November 9, 1922, p. 724.

Lea, K. M. *Italian Popular Comedy: A Study in the Commedia Dell' Arte, 1560-1620 With Special Reference to the English Stage.* 2 vols. Oxford, 1934.

McKerrow, Ronald B. *An Introduction to Bibliography for Literary Students.* Oxford, 1927.

————. "A Suggestion Regarding Shakespeare's Manuscripts," RES, XI (1935), 459-465.

Matthews, William. "Shakespeare and the Reporters," *Lib.*, 4th Ser., XV (1935), 481-498.

————. "Shorthand and the Bad Shakespeare Quartos," MLR, XXVII (1932), 243-262.

Mattingly, Garrett. *Renaissance Diplomacy.* Boston, 1955.

Meadley, T. D. "Attack on the Theatre (*circa* 1580-1680)," *The London Quarterly and Holborn Review* (January 1953), pp. 36-41.

Morris, J. E. "The Date of 'Henry IV,' " TLS, January 28, 1926, p. 62.

Muir, Kenneth. *Shakespeare's Sources: Comedies and Tragedies.* I. London, 1957.

Munro, John. "Some Matters Shakespearean—I," TLS, September 13, 1947, p. 472.

Murray, John Tucker. *English Dramatic Companies 1588-1642.* 2 vols. London, 1910.

Neale, J. E. *Queen Elizabeth.* London and New York, 1934.

Nicolas, Nicholas Harris. *History of the Orders of Knighthood of the British Empire.* . . . 2 vols. London, 1841.

————. *Memoir of Augustine Vincent, Windsor Herald.* London, 1827.

Nosworthy, J. M. "The Merry Wives of Windsor," TLS, September 27, 1947, p. 497.

Ogburn, Vincent H. "The Merry Wives Quarto, a Farce Interlude," PMLA, LVII (September 1942), 654-660.

Palme, Per. *Triumph of Peace*. Stockholm, 1956.

Parrott, Thomas Marc. *Shakespearean Comedy*. New York,
1949.

Pollard, Alfred W. "Elizabethan Spelling as a Literary and
Bibliographical Clue," *Lib.*, 4th Ser., IV (June 1, 1923), 2-8.

——. "The Foundations of Shakespeare's Text," *Aspects of
Shakespeare*, pp. 1-22. British Academy Lecture, April 23,
1923. Oxford, 1933.

——. *Shakespeare Folios and Quartos: A Study in the Bibliography of Shakespeare's Plays, 1594-1685*. London, 1909.

——. *Shakespeare's Fight with the Pirates and the Problems
of the Transmission of his Text*, 2nd ed. rev. Cambridge,
England, 1920.

Pollard, Alfred W. and J. Dover Wilson. "The 'Stolne and
Surreptitious' Shakespearian Texts," TLS, January 9, 1919,
p. 18 (I. "Why Some of Shakespeare's Plays were Pirated");
January 16, 1919, p. 30 (II. "How Some of Shakespeare's
Plays were Pirated"); March 13, 1919, p. 134 ("Henry V");
August 7, 1919, p. 420 ("The Merry Wives of Windsor"
[1602]); August 13, 1919, p. 434 ("Romeo and Juliet").

Pote, Joseph. *The History and Antiquities of Windsor Castle
. . . .* Eton, 1749.

Rhodes, R. Crompton. "The Early Editions of Sheridan," TLS,
September 17, 1925, p. 599 ("The Duenna"); September 24,
1925, p. 617 ("The School for Scandal").

——. *Shakespeare's First Folio*. Oxford, 1923.

——. "Some Aspects of Sheridan Bibliography," *Lib.*, 4th
Ser., IX (1928), 233-261.

Robertson, J. M. "The Problem of 'The Merry Wives of Windsor,'" *Shakespeare Association Papers*, No. 2, November 9,
1917.

Robson-Scott, W. D. *German Travellers in England 1400-1800*.
Oxford, 1953.

Rosenberg, S. L. Millard. "Duke Friedrich of Württemberg,"
SAB, VIII (April 1933), 92-93.

——. "The Original of the 'Duke de Jarmany,'" *University
of California Chronicle*, XXXV (January 1933), 90-93.

Sarrazin, Gregor. "Nym und Ben Jonson," *Jahrbuch der Deutschen Shakespeare Gesellschaft*, XL (1904), 212-222.

Schell, J. Stewart. "Shakespeare's Gulls," SAB, XV (January 1940), 23-33.

Schücking, Levin L. "The Fairy Scene in 'The Merry Wives' in Folio and Quarto," MLR, XIX (July 1924), 338-340.

———. "The Quarto of King Henry IV., Part II," TLS, September 25, 1930, p. 752.

Sewell, Sallie. "The Relation between *The Merry Wives of Windsor* and Jonson's *Every Man in His Humour*," SAB, XVI (July 1941), 175-189.

Shaw, William A. *The Knights of England*. 2 vols. London, 1906.

Simpson, Percy, ed. *Ben Jonson's Every Man in his Humour*. Oxford, 1919 (reprinted 1936).

Sisson, C. J. *New Readings in Shakespeare*. 2 vols. Cambridge, 1956.

———. "Shakespeare's Quartos as Prompt-Copies: with some account of Cholmeley's Players and a new Shakespeare Allusion," RES, XVIII (April 1942), 129-143.

Stamp, A. E. *The Disputed Revels Accounts*. Oxford, 1930.

Steele, Mary Susan. *Plays & Masques at Court during the Reigns of Elizabeth, James, and Charles*. New Haven, Conn., and London, 1926.

Stender, John L. "Master Doctor Caius," *Bulletin of the History of Medicine*, VIII (January 1940), 133-138.

Stopes, C. C. *Burbage and Shakespeare's Stage*. London, 1913.

———. *Shakespeare's Warwickshire Contemporaries*. Stratford-Upon-Avon, 1907.

———. *William Hunnis and the Revels of the Chapel Royal*. London, 1910.

Tannenbaum, Samuel. "Prof. Hotson's Conclusions about Shakespeare Disputed," *New York Times*, October 18, 1931, Sec. 3, p. 2.

Taylor, George C. "Did Shakespeare, Actor, Improvise in Every Man in his Humour?" *Joseph Quincy Adams Memorial Studies*, ed. James G. McManaway and others, pp. 21-32. Washington, D.C., 1948.

Thomas, S. G. "Source of 'The Merry Wives of Windsor,'" TLS, October 11, 1947, p. 528.

Tighe, Robert Richard and James Edward Davis. *Annals of Windsor, being a History of the Castle and Town.* . . . 2 vols. London, 1858.

Walker, Alice. *Textual Problems of the First Folio.* Cambridge, England, 1953.

Wallace, Charles W. *The Children of the Chapel at Blackfriars, 1597-1603.* University Studies of the University of Nebraska, VIII. Lincoln, 1908.

Weaver, F. J. "Anglo-French Diplomatic Relations, 1558-1603," *Bulletin of the Institute of Historical Research*, IV (November 1926), 73-86; V (June 1927), 13-22; VI (June 1928), 1-9; VII (June 1929), 13-26.

Wentersdorf, Karl. "Shakespearean Chronology and the Metrical Tests," *Shakespeare-Studien: Festschrift für Heinrich Mutschmann*, pp. 161-193. Marburg, 1951.

White, David Manning. "An Explanation of the *Brooke-Broome* Question in Shakespeare's *Merry Wives*," PQ, XXV (July 1946), 280-283.

———. "The Textual History of 'The Merry Wives of Windsor.'" Unpublished dissertation. University of Iowa, 1942.

Whittaker, Herbert. "Full Shakespeare Texts Return—with Bonuses," *Toronto Globe and Mail*, August 11, 1956, p. 22.

Willoughby, Edwin Eliot. *A Printer of Shakespeare: The Books and Times of William Jaggard.* London, 1934.

Wilson, F. P. "Ralph Crane, Scrivener to the King's Players," *Lib.*, 4th Ser., VII (1926), 194-215.

Wilson, J. Dover. *The Essential Shakespeare: A Biographical Adventure.* Cambridge, England, 1937.

———. "The Origins and Development of Shakespeare's Henry IV," *Lib.*, 4th Ser., XXVI (June 1945), 2-16.

Withington, Robert. *English Pageantry, an Historical Outline.* 2 vols. Cambridge, Mass., and London, 1918 and 1920.

INDEX ✦

In describing subject modifications, the following shortened forms have been used: Duke Frederick = Frederick, Duke of Württemberg; Garter = Order of the Garter; *1 HIV* = *Henry IV*, part 1; *2 HIV* = *Henry IV*, part 2; *HV* = *Henry V*; *MWW* = *The Merry Wives of Windsor*.

abridgment of plays, 79, 81-83
achievements (heraldry), 11, 44
Act to Restrain Abuses of Players, 101
acting companies, provincial, 83
Allen, Don Cameron, 209, 210
Anglo-French diplomatic relations, 1596, 139n37, 167, 168
Anglo-German trade relations, 1595-1604, 97, 132, 133, 135, 138-46, 173. *See also* Frederick, Duke of Württemberg
Anglo-Hanse relations, 1595-1604, *see* Anglo-German trade relations
Annals (Camden), 133
Annals (Stow), 211, 212
Anne, Queen (wife of James I), 174
Ascham, Roger, 210
Ashmole, Elias, 26, 29, 30, 46, 138n33, 139n37, 171
"assembled text" theory, 102n
Atkinson, Brooks, 184n11

Bacon, Anthony, 56, 57, 63n4
Bacon, Francis, 56, 63n4, 182
"bad" quartos, 3, 78n12, 92, 100, 101
Badenfahrt . . . , 141
Baldwin, T. W., 98
Bardolph, 107, 121, 154, 155, 170, 175, 189, 190
Bedford, Earl of, 12, 13, 15
Berkeley, Sir Richard, 30n17
Black Book, 46, 47
Blind Beggar of Alexandria, The, 203
Blue Book (Garter annals), 31n18, 33, 34, 35, 36, 46, 55, 56n7, 134, 135
Bouillon, Duke of, 37, 167
Brackenbury, Richard, 38, 64
Bracy, William, 73n2, 75n4, 81, 82, 87, 110n, 203n14, 204

Braun, George, 8
Breuning von Buchenbach, Hans Jacob, 27, 34, 137n32, 143; Garter mission to England, 125-32
Brittania . . . (Camden), 8
Brock, Elizabeth, 75n4, 104n62, 153n
Brooke-Broome name variant, 5. *See also* Brooke, Master
Brooke, George, role in the Bye Plot, 110-11
Brooke, Henry, Lord Cobham, *see* Cobham, Henry Brooke, Lord
Brooke-Jaggard-Vincent quarrel, 108-09
Brooke, Master, variant reading for Broome, 107-20
Brooke, Ralph, York Herald, 108-09, 111
Brooke, William, Lord Cobham, *see* Cobham, William Brooke, Lord
Broome, Master, variant reading for Brooke, 107-20
Browne, Sir William, 111
Buckhurst, Lord, 43, 44
Burbage, Richard, 98n51
Burgh, Lord Thomas, 14, 32, 45
Burghley, Thomas Cecil, Lord, 38, 56
Burghley, William Cecil, Lord, 30n17, 126, 127, 129, 137, 140, 167
Busby, John, 75, 92, 93, 99, 100
Buwinckhausen von Wallmerode, Benjamin, 125, 143, 144n50, 145, 146, 150n
Bye Plot, 110

Cadiz, 167, 179, 181
Caius, Dr., 12, 17, 116, 196, 197, 204; as choleric Frenchman, 205-06; and the "grand affair," 10, 11, 12, 16, 30; in horse-stealing subplot,

[231]